Inverse and Ill-Posed Problems Series

Integral Geometry and Inverse Problems for Kinetic Equations

Also available in the Inverse and Ill-Posed Problems Series:

Computer Modelling in Tomography and Ill-Posed Problems
M.M. Lavrent'ev, S.M. Zerkal and O.E. Trofimov
An Introduction to Identification Problems via Functional Analysis
A. Lorenzi
Coefficient Inverse Problems for Parabolic Type Equations and Their Application
P.G. Danilaev
Inverse Problems for Kinetic and other Evolution Equations
Yu.E. Anikonov
Inverse Problems of Wave Processes
A.S. Blagoveshchenskii
Uniqueness Problems for Degenerating Equations and Nonclassical Problems
S.P. Shishatskii, A. Asanov and E.R. Atamanov
Uniqueness Questions in Reconstruction of Multidimensional Tomography-Type Projection Data
V.P. Golubyatnikov
Monte Carlo Method for Solving Inverse Problems of Radiation Transfer
V.S. Antyufeev
Introduction to the Theory of Inverse Problems
A.L. Bukhgeim
Identification Problems of Wave Phenomena - Theory and Numerics
S.I. Kabanikhin and A. Lorenzi
Inverse Problems of Electromagnetic Geophysical Fields
P.S. Martyshko
Composite Type Equations and Inverse Problems
A.I. Kozhanov
Inverse Problems of Vibrational Spectroscopy
A.G. Yagola, I.V. Kochikov, G.M. Kuramshina and Yu.A. Pentin
Elements of the Theory of Inverse Problems
A.M. Denisov
Volterra Equations and Inverse Problems
A.L. Bughgeim
Small Parameter Method in Multidimensional Inverse Problems
A.S. Barashkov
Regularization, Uniqueness and Existence of Volterra Equations of the First Kind
A. Asanov
Methods for Solution of Nonlinear Operator Equations
V.P. Tanana
Inverse and Ill-Posed Sources Problems
Yu.E. Anikonov, B.A. Bubnov and G.N. Erokhin
Methods for Solving Operator Equations
V.P. Tanana
Nonclassical and Inverse Problems for Pseudoparabolic Equations
A. Asanov and E.R. Atamanov
Formulas in Inverse and Ill-Posed Problems
Yu.E. Anikonov
Inverse Logarithmic Potential Problem
V.G. Cherednichenko
Multidimensional Inverse and Ill-Posed Problems for Differential Equations
Yu.E. Anikonov
Ill-Posed Problems with A Priori Information
V.V. Vasin and A.L. Ageev
Integral Geometry of Tensor Fields
V.A. Sharafutdinov
Inverse Problems for Maxwell's Equations
V.G. Romanov and S.I. Kabanikhin

INVERSE AND ILL-POSED PROBLEMS SERIES

Integral Geometry and Inverse Problems for Kinetic Equations

A.Kh. Amirov

///VSP///

UTRECHT • BOSTON • KÖLN • TOKYO

2001

VSP
P.O. Box 346
3700 AH Zeist
The Netherlands

Tel: +31 30 692 5790
Fax: +31 30 693 2081
vsppub@compuserve.com
www.vsppub.com

First published in 2001

ISBN 90-6764-352-1

Printed in The Netherlands by Ridderprint bv, Ridderkerk.

Abstract

A new method for proving the solvability of problems of integral geometry and inverse problems for kinetic equations is presented. Application of this method has lead to interesting problems of Dirichlet type for third order partial differential equations. It appeared that the solvability of the latter essentially depends on the geometry of the domain for which the problem is stated. Another subject considered in the book concerns the problem of integral geometry on paraboloids. In particular, we investigate the uniqueness of solutions to the Goursat problem for a differential inequality, which implies new theorems on the uniqueness of solutions to this problem for a class of quasilinear hyperbolic equations. The scope of the book also includes a class of multidimensional inverse problems associated with problems of integral geometry and the inverse problem for the quantum kinetic equation.

The monograph is intended for the mathematicians who deal with problems of integral geometry, direct and inverse problems of mathematical physics and geophysics, for specialists in computerized tomography, and for graduating and postgraduate students.

Bibliography: 129 titles.

Contents

Introduction

Problems of integral geometry are defined as follows. Let $\lambda(x)$ be a sufficiently smooth function defined in n-dimensional space, $x = (x_1, \dots, x_n)$, and assume that $\{\Gamma(r)\}$ is a family of smooth manifolds in this space which depend on the parameter $r = (r_1, \dots, r_k)$. Suppose that the integrals

$$\int_{\Gamma(r)} \lambda(x)\, d\sigma = J(r), \tag{1}$$

are known, where $d\sigma$ is the measure element on $\Gamma(r)$. Given $J(r)$, it is required to find the function $\lambda(x)$.

The following questions arise in the study of this problem:

1. Is $\lambda(x)$ uniquely determined by $J(r)$?

2. How can $\lambda(x)$ be expressed in terms of $J(r)$ by means of an analytic formula? We note that only a few exceptional cases are possible.

3. What are the necessary and sufficient conditions for $J(r)$ to belong to the class of functions that can be represented in the form (1)?

In this book, the problems of integral geometry are treated in the case of one-dimensional manifolds $\Gamma(r)$, or more precisely, in the case where $\Gamma(r)$ are curves and $d\sigma$ is the arc length element of a curve. Historically, the Radon transform (see Radon, 1917) is likely to be related to the first problem of integral geometry. The Radon transform maps the function $\lambda(x)$ to the integrals of λ over different hyperplanes. Various aspects of this problem were considered later by John (1935, 1943), Kostelyanets and Reshetnyak (1954), Gel'fand and Graev (1959), Gel'fand, Graev, and Vilenkin (1962), and some other researchers. These problems are closely related to the theory of Lie group representations (see Gel'fand, Graev, and Vilenkin, 1962). Other important results can be found in Courant (1964), John (1955), Plaksin (1966),

Romanov (1972, 1978, 1984), Santalo (1976), Uspenskii (1972, 1977), and Helgason (1980).

The theory of problems of integral geometry was developed further in Amirov (1986, 1987), Anikonov (1976, 1978, 1987, 1997), Kireitov (1975, 1983), Lavrent'ev and Bukhgeim (1973), Mukhometov (1975a, 1975b, 1977, 1981), Mukhometov and Romanov (1978), Romanov (1978), Lavrent'ev, Derevtsov, and Sharafutdinov (1977), Pestov and Sharafutdinov (1988), Sharafutdinov (1981, 1994a, 1994b, 1993), and other works.

Inverse problems are problems of determining coefficients, the right-hand side, initial conditions or boundary conditions of a differential equation from some additional information about a solution of the equation. Such problems appear in many important applications of physics, geophysics, technology, and medicine.

One of the characteristic features of inverse problems for differential equations is their being ill-posed in the sense of Hadamard. The general theory of ill-posed problems and their applications is developed by A. N. Tikhonov, V. K. Ivanov, M. M. Lavrent'ev, and their pupils (see Anikonov, 1978; Ivanov, Vasin, and Tanana, 1978; Lavrent'ev, 1976; Lavrent'ev, Romanov, and Shishatskii, 1980; Morozov, 1974; Romanov, 1984; Tanana, 1997; Tikhonov and Arsenin, 1979; Tikhonov, Arsenin, and Timonov, 1987).

Problems of integral geometry for complex manifolds are connected with inverse problems for differential equations. This fact was first pointed out in Lavrent'ev and Romanov (1966). The connection was used later in Amirov (1977, 1986, 1987a, 1987d), Blagoveshchenskii (1990), Bukhgeim (1972), Golubyatnikov (1995), Klibanov (1976a, 1976b), Lavrent'ev and Amirov (1976, 1977a, 1977b, 1977c), Lavrent'ev and Anikonov (1967), Mukhometov (1975a, 1975b, 1977, 1981), Mukhometov and Romanov (1978), and Romanov (1978).

Inverse problems for kinetic equations appear to be important both from theoretical and practical points of view. Interesting results in this field are presented in Amirov (1983, 1984a, 1984b, 1985, 1986a, 1986b, 1986c, 1986d, 1986e, 1987a, 1987b, 1987c, 1987d), Amirov and Pashaev (1992), Anikonov (1984a, 1984b, 1978, 1987, 1997), Anikonov and Amirov (1983), Anikonov and Bondarenko (1982, 1984), Kabanikhin and Boboev (1981), Kireitov (1983), Lavrent'ev and Anikonov (1967), Pestov and Sharafutdinov (1988), Prilepko and Ivankov (1985), Prilepko and Volkov (1987), Romanov, Kabanikhin, and Boboev (1983), Hamaker, Smith, Solmon, and Wagner (1980). The physical meaning of these problems consists in finding particle interaction forces, scattering indicatrices, radiation sources and other physical quantities.

Integral geometry and inverse problems for kinetic equations are closely interrelated: many problems of integral geometry are equivalent to the corresponding inverse problems for kinetic equations, and vice versa (see Appendix A, Part 1).

The problems of determining isotropic sources were prevalent in earlier works on this subject. In this book, we study a new class of problems of integral geometry and inverse problems for kinetic equations, in which the sources can essentially depend on the speed. This makes it possible to extend the classes of uniqueness of solutions and, in some cases, to prove the existence theorems for the corresponding problems, in particular, for the problem of integral geometry with isotropic sources.

The first uniqueness theorem in this direction was proved by Anikonov and Amirov (1983).

At present, there are a great number of publications on the uniqueness of solutions to problems of integral geometry and inverse problems for kinetic equations, while the problem of existence has been given much less attention. Concerning the latter, it is worth mentioning the classical work by Radon (1917), the works by Romanov (1972), Anikonov (1976), and Bukhgeim (1983).

At the same time, the existence theorems for these problems are of great interest from the point of view of both theory and applications.

The main difficulty in establishing the solvability of such problems is that a problem of integral geometry (or the equivalent inverse problem for the kinetic equation) is overdetermined if the sources depend only on x ($\lambda = \lambda(x)$). Therefore, the initial data of such a problem should not be arbitrary; they should satisfy nontrivial "solvability conditions". These conditions are extremely hard to find. For example, as follows from the well-known results for the Radon transform (see Gel'fand, Graev, and Vilenkin, 1962; Ludwig, 1966; Tikhonov, Arsenin, and Timonov, 1987), a problem of integral geometry along straight lines has a solution in a domain D if and only if the so-called "Cavalieri condition" holds. Before formulating these conditions, we introduce the following notation: $Y = S^1 \times R$, where S^1 is the unit circle in $\mathbb{R}^2$ centred at the origin; $r = (w, s)$ represents the parameters of the line $\Gamma(r) = \{x : \langle x, w\rangle = s\}$;

$$J(w,s) = \begin{cases} u_0(x^{(2)},\varphi) - u_0(x^{(1)},\varphi), & \text{if } \{\langle x, w\rangle = s\} \cap \partial D = \{x^{(1)}, x^{(2)}\} \\ 0, & \text{otherwise} \end{cases}$$

$x^{(1)}$ and $x^{(2)}$ are the points of intersection of the line $\langle x, w\rangle = s$ and the boundary of D, $u_0(x, \varphi)$ is the trace of $u(x, \varphi)$ on $\Gamma_1 = \partial D \times (0, 2\pi)$, and $u(x, \varphi)$ is a solution to Problem 1.1.1 with $f = 0$. In this case, Problem 1.1.1 is equivalent to the above problem of integral geometry along straight lines (see p. 1 and Remark 1.1.3). Let $\operatorname{supp} \lambda \subset \mathrm{D}$. The "Cavalieri condition" is formulated as follows: The function $J(w, s) \in H^{1/2}(Y)$ can be represented in the form (1) for $\lambda(x) \in L_2(D)$ if and only if

(a) $J(w, s)$ is a finite function,

(b) $J(w, s) = J(-w, s)$,

(c) for any natural m, $\displaystyle\int_{\mathbb{R}} s^m J(w, s)\, ds$ is a polynomial with respect to w of degree not higher than m.

Note that the above formulation of the "Cavalieri condition" might seem ill posed because it is required that $\lambda(x)$ be integrable over the lines $\Gamma(r)$, whereas the trace of $\lambda(x)$ is, in general, undefined on $\Gamma(r)$ since the planar measure of $\Gamma(r)$ is zero. However, in fact, everything is well defined here because Fubini's theorem implies that for any $x \in D$ and almost all $\varphi \in (0, 2\pi)$ the function $\lambda(x)$ has a trace on the line $\Gamma(r)$ passing through the point x in the direction $(\sin \varphi, \cos \varphi)$, and this trace belongs to $L_2(\Gamma(r))$ (see Mikhailov, 1976).

It should be noted that the set of functions u_0 for which the problem of integral geometry is solvable is not everywhere dense in any of the spaces $L_2(\Gamma_1)$, $C^m(\Gamma_1)$, and $H^m(\Gamma_1)$.

As a rule, the data in problems of integral geometry are of quasianalytic character, i.e., their values specified in a domain of Lebesgue measure as small as desired determine their values in an essentially larger domain (see Courant, 1964; Romanov, 1972). In particular, this implies that it's impossible to avoid overdetermination by specifying the data on a part of the boundary rather than on the whole boundary. Even if it were possible to find the solvability conditions for the mentioned overdetermined problems, it seems that these conditions would not always be satisfactory for those who deal with applications. The reason is that real data in practice usually have some errors and thus fall out of the data class for which the existence of a solution is established.

We outline the general procedure of the method for establishing the solvability of a problem of integral geometry.

An overdetermined problem is replaced by a determined one as follows: The unknown function λ in the problem of integral geometry is assumed to depend not only on x but also on the direction ξ (or φ) in a specific way, that is, we have $\lambda = \lambda(x, \xi)$. (The dependence on ξ cannot be arbitrary because the problem is underdetermined in this case. Examples of multiple solutions can be easily constructed.) Specific dependence of λ on the direction means that λ is a solution of a certain differential equation with the following two properties:

1. the problem of integral geometry with $\lambda = \lambda(x, \xi)$ is determined;

2. the functions that depend only on x satisfy this equation.

In general, the above-mentioned equation is not uniquely determined for a specific problem.

Property 1 means that the class of the unknown functions λ is extended so that the problem of integral geometry becomes well posed for the new class.

Property 2 indicates that the above extension is not arbitrary: it should contain the functions depending only on x (as in classical problems of integral geometry).

Suppose that we have found a differential equation for λ that satisfies properties 1 and 2. Suppose also that we know *a priori* that the function u_0^e represents the exact data of the problem of integral geometry for the function $\lambda = \lambda(x)$. Then we can construct a solution $\tilde{\lambda}$ of the problem of integral geometry. By uniqueness, $\tilde{\lambda}$ and $\lambda(x)$ coincide. At the same time, knowing the approximate data u_0^a such that $\|u_0^e - u_0^a\|_{H^3(\Gamma_1)} \leq \varepsilon$, we can construct an approximate solution λ^a such that $\|\lambda - \lambda^a\| \leq C\varepsilon$. Here the data are specified on Γ_1 and $C > 0$ is independent of u_0^e and u_0^a.

In other words, we have a regularizing procedure for the problem under consideration. This procedure is used in the first chapter of this book to study the solvability of a class of problems of integral geometry, whereas in the second chapter (Section 2.2) it is used to study an inverse problem for the kinetic equation. For each of these problems, we construct an equation with properties 1 and 2 and a space in which it has a unique solution, this space being essentially dependent on the problem. In both cases, however, certain spaces are given that serve as lower and upper "bounds", respectively.

The uniqueness theorems presented in this book are proved mostly with the use of *a priori* estimates (in some cases, with weight functions of the Carleman type).

Application of the above method of proving the existence theorem to the problems of integral geometry and inverse problems for kinetic equations lead to interesting problems, such as problems of the Dirichlet type for the third order equation of the form

$$Au \equiv \widehat{L}Lu = 0, \tag{2}$$

where $\widehat{L}$ and L are second and first order differential expressions, respectively, defined in the domain Ω.

For the problems corresponding to the problems of integral geometry from Sections 1.1 and 1.4, the data for the solution of equation (2) are specified on a part of the boundary of the domain Ω: $\Gamma_1 = \partial D \times (0, 2\pi)$ and $\Gamma_1 = \partial D \times (0, L)$. It is required that the solution be periodic in φ or z.

For the problems associated with the problems from Sections 1.3 and 2.2, the data for the solution of (2) are specified on the whole boundary of Ω. It should be noted that the geometry of the domain Ω is essential for the solvability of these problems. More precisely, it is important that Ω can be represented in the form of the direct product of two domains in the spaces of x and ξ (or φ), correspondingly. These requirements are illustrated by several examples given in the book (see Remarks 1.3.2 and 2.2.1).

Note that equation (2) is written for generalized functions, and solutions of problems of Dirichlet type for (2) are sought in the appropriate classes of generalized functions.

Before providing a brief outline of the contents of the present work, we consider the following problem of integral geometry (see Bukhgeim, 1972).

It is required to determine a function $\lambda(x, y)$ from its integrals over a set of paraboloids, that is, knowing the function

$$f(x,y) = \int_{S(x,y)} \lambda(\xi, \eta) d\xi, \qquad S(x,y) : \ \eta = y - \langle A(x-\xi), x-\xi \rangle,$$

where A is a positive Hermitian matrix. We shall restrict ourselves to the case where A is the unit matrix, since the general case can be reduced to it by a linear change of variables. Thus, we have $S(x, y)$: $\eta = y - |x - \xi|^2$. Then the problem can be reformulated as follows: given the function

$$f(x,y) = \int_{|x-\xi| \le \sqrt{y}} \lambda(\xi, y - |x-\xi|^2) d\xi,$$

find $\lambda(x, y)$.

Assume

$$u(x,y,t) = t^{-n/2} \int_{|x-\xi| \le \sqrt{yt}} \lambda(\xi, y - t^{-1}|x-\xi|^2) d\xi, \qquad t > 0.$$

Then, if $\lambda \in C^\infty(\mathbb{R}_+)$ and $\lambda(x,0) = \lambda'_y(x,0) = 0$, then $u(x,y,t)$ is infinitely differentiable for $y > 0$, $t > 0$, and the following relations hold:

$$L_1[u] = \left(4\frac{\partial^2}{\partial y \partial t} - \Delta_x\right)u = 0, \qquad (x,y) \in \mathbb{R}_+,\ t > 0 \tag{3}$$

$$u(x,y,1) = f(x,y), \qquad u(x,0,t) = 0,$$

$$u(x,y,0) = \frac{\omega_n}{2}\int_0^y \lambda(x,s)(y-s)^{n/2-1}ds.$$

Here ω_n is the area of the unit sphere in $\mathbb{R}^n$ and Δ_x is the Laplacian with respect to x.

The structure of the book is outlined as follows. At the beginning of Chapter 1 (Section 1.1) we study a planar problem of integral geometry. The three-dimensional analog of this problem is considered in Section 1.2. Section 1.3 is devoted to the problem of integral geometry along geodesics of a Riemannian space. Section 1.4 deals with the solvability of a planar problem of integral geometry. The uniqueness of solution of this problem for the unknown functions that depend only on x was established by Mukhometov (1975) for the case of $C^2(\Omega)$. Certain problems of tomography are considered in Section 1.5.

Chapter 2 contains mostly the proofs of the uniqueness theorems for multidimensional problems for kinetic equations: the Liouville equation, the Boltzmann equation, and the Vlasov system. The existence of solutions is established for several problems (Sections 2.2 and 2.4).

In Chapter 3, boundary value problems for evolutionary equations associated with equation (3) are studied. The first two sections (3.1 and 3.2) concern the well-posedness of the Cauchy problem for the equation

$$\frac{\partial^m u}{\partial t^m} = a\int_0^y \Delta_x^p u(x,\tau,t)d\tau.$$

We investigate how this problem being well posed or ill posed depends on different combinations of (m,p) and the sign of a; examples of the Hadamard type for ill-posed problems are constructed.

Section 3.3 deals with the Goursat problem for a differential inequality associated with equation (3). The results of this section, in particular, imply new theorems of uniqueness "in the small" for the Goursat problem for a class of quasilinear hyperbolic equations. Section 3.4 is devoted to the study of the conditional well-posedness of the Cauchy problem for the abstract analog of equation (3) in a Hilbert space.

In Section 3.5, the solvability of the inverse problem for the evolutionary equation from Section 3.4 is studied.

The uniqueness theorems for a class of inverse problems for second order differential equations are proved in Chapter 4.

Section 4.1 is devoted to the study of the inverse problem for the quantum kinetic equation. The Fourier transform with respect to p (momenta) of the quantum kinetic equation leads to the inverse problem for the corresponding ultrahyperbolic equation. Section 4.2 is devoted to the study of the inverse problem for an ultrahyperbolic equation of more general form.

In Section 4.3 we consider the inverse problem for a hyperbolic equation (special case of the corresponding equation from Section 4.2). These problems are known to be related to integral geometry (see Romanov, 1972). Also, inverse problems for a class of parabolic and elliptic equations are analyzed. The distinguishing feature of the problem statements considered here, as compared to other studies on this subject, is that the data of the direct problem in Section 4.3 are specified on a planar part of the domain boundary. In some cases this part may have arbitrarily small measure. The problem studied in Section 4.4 is associated with the inverse problem for a hyperbolic equation whose data involve the Dirac function. Such problems are connected with problems of integral geometry on ellipsoids. Besides, they are of great interest, for example, for geophysicists. The reason is that in applications related to the problems of exploration geophysics the data are often represented by finite functions vanishing outside a relatively small domain. In mathematical formalizations of these problems in such situations, the data are represented using generalized functions with supports concentrated at fixed points of the corresponding space (see Romanov, 1984). The uniqueness of solutions of the problems considered in Section 3.3 and Chapter 4 is established with the use of the Carleman method. A specific weight function of the Carleman type is constructed for each problem.

The most important kinetic equations are presented in Appendix A.

We point out some applications of problems of integral geometry and inverse problems for kinetic equations. First, these problems are applied in some fields of pure mathematics, namely,

(a) the theory of Lie group representations (see Gel'fand and Graev, 1959),

(b) the study of coefficient inverse problems for hyperbolic equations (see Romanov, 1984),

(c) inverse problems of spectral analysis (see Amirov and Pashaev, 1992; Belishev, 1988),

(d) the problem of determining a Riemannian metric from the distances between boundary points of a domain (see Bernshtein and Gerver, 1978, 1980; Gerver and Markushevich, 1967; Mukhometov, 1977; Mukhometov and Romanov 1978).

The mentioned problems also have practical applications. For example, they provide mathematical background for computerized tomography (see Natterer, 1986). The operation principle of computerized tomography consists in the following (see Tikhonov, Arsenin, and Timonov, 1987). The object under investigation is exposed to radiation at different angles, and the radiation parameters are measured at the points of observation. The results are digitized and processed by computers which calculate spatial distribution of quantitative physical parameters of the object. The obtained results are then visualized by means of special devices.

The basic equation in the mathematical model of computerized tomography can be written in the general form of equation (1), where $\Gamma(r)$ is the ray connecting the source point of the object radiation with the observation point; $\lambda(x)$ characterizes the object under study.

Examples of important applications of computerized tomography are given below.

(a) X-ray tomographs are most commonly employed in modern medicine. As distinct from radiographs, tomograms have no shadows and it is possible to adjust the contrast and brightness of images on the display.

It is well known that computerized tomography caused a revolution in medical diagnostics and electron-microscope investigation of biological macromolecules. The inventors of the diagnostic technique of computerized tomography, A. Cormack and G. N. Hounsfield, and the researcher who pioneered in applying this technique in biochemistry, A. Klug, were awarded the Nobel prizes in 1979 and 1982, respectively.

(b) Computerized tomography is also applied in geophysics. Instantaneous point explosions can be used as the radiation sources (x^0). Suppose that the propagation of waves takes place in a domain $D \subset \mathbb{R}^3$ with boundary S, with finite speed $v(x)$, $x = (x_1, x_2, x_3)$ (see Romanov 1984). Let $\tau(x, x^0)$ be the time required for the wave coming from the source x^0 to reach the observation point x. Given the function $\tau(x, x^0)$ $(x, x^0 \in S)$, it is required to find $v(x)$ ($v(x) = |\nabla_x \tau(x, x^0)|$, ∇_x is the gradient with respect to x). Note that $\tau(x, x^0)$ and $v(x)$ are related to $J(r)$ and $\lambda(x)$ from equation (1): $\tau(x, x^0) = J(r)$ and $\lambda(x) = 1/v(x)$.

(c) Computerized tomography is extensively put into practice in flaw detection and microscopy. It is not only X-radiation that is used in investigations, but also gamma-radiation, various heavy particles and electrons, long-wave electromagnetic radiation, etc. Tomographic technique of flaw detection plays an important role in the inspection of hard-to-reach places of various structures.

Computerized tomography is also effective in microscopy because the appropriate focusing of radiation on the examined surface often cannot be achieved with existing equipment. In practice, this problem is illustrated by shadow effects in optical-microscope and electron-microscope investigations. A comprehensive survey of application of computerized tomography in microscopy is presented in Vainshtein (1973).

There are a number of other important applications of computerized tomography, which are not mentioned here.

We introduce the notation that will be used in the sequel.

$\mathbb{R}^n$ denotes the n-dimensional Euclidean space; for $x \in \mathbb{R}^n$, where $x = (x^1, x^2, \dots, x^n)$, set $'x = (x^2, x^3, \dots, x^n)$ and $x' = (x^1, x^2, \dots, x^{n-1})$. For a bounded domain G, $C^m(G)$ is the Banach space of functions that are m times continuously differentiable in G; $C^\infty(G)$ is the set of functions that belong to $C^m(G)$ for all $m \geq 0$; $C_0^\infty(G)$ is the set of finite functions in G that belong to $C^\infty(G)$; $L_2(G)$ is the space of measurable functions that are square integrable in G, $H^k(G)$ is the Sobolev space and $\overset{\circ}{H}{}^k(G)$ is the closure of $C_0^\infty(G)$ with respect to the norm of $H^k(G)$.

These standard spaces are described in detail, for example, in Lions and Magenes (1972) and Mikhailov (1976). The definitions of other special spaces that will be used are given in the corresponding sections of this book.

I wish to thank M. M. Lavrent'ev, who familiarized me with the theory of ill-posed problems, for his constant support and for useful discussions of the problems analyzed in this book. I am grateful to Yu. E. Anikonov, who encouraged me to study inverse problems for kinetic equations, for his numerous helpful advice on this subject. I also wish to thank my colleagues from the Department of Conditionally Well-Posed Problems, particularly those from the Laboratory of Inverse Problems of Mathematical Physics, the Sobolev Institute of Mathematics, Siberian Branch of the Russian Academy of Sciences, for warm sympathy and cooperation in research during my work at the Department of Conditionally Well-Posed Problems.

Chapter 1.

Solvability of problems of integral geometry

One of the main methods proposed here for investigating the solvability of problems of integral geometry is to reduce them to equivalent Dirichlet problems for differential equations. Such a reduction is demonstrated for Problem A formulated below (see also (1.3.6), (1.4.4), (1.4.5)). For reference, see Amirov (1984a, 1986a), Lavrent'ev and Anikonov (1967), and Mukhometov (1975).

Problem *A*. Find a function $\lambda(x)$ in a domain D from the integrals of λ along the curves of a given family of curves $\{\Gamma\}$. It is assumed that for any point $x \in D$ and any direction $\nu = (\sin\varphi, \cos\varphi)$ there exists a unique sufficiently smooth curve in $\{\Gamma\}$, with endpoints on the boundary of D, passing through the point x in the direction ν.

Suppose that $\lambda(x) \in C(\mathbb{R}^2)$ vanishes outside D. Introduce an auxiliary function

$$u(x,\varphi) = \int_{\gamma(x,\varphi)} \lambda \, ds, \tag{1.0.1}$$

where the curve $\gamma(x,\varphi)$ passes through x in the direction ν and is a part of a curve with the same property that belongs to Γ. One end of $\gamma(x,\varphi)$ is the point x and the other one lies on the boundary of D. Differentiating (1.0.1) in the direction tangent to $\gamma(x,\varphi)$ at x, we get the equation

$$u_{x_1} \sin\varphi + u_{x_2} \cos\varphi + u_\varphi f(x,\varphi) = \lambda(x), \tag{1.0.2}$$

where $f(x, \varphi)$ is the curvature of $\gamma(x, \varphi)$ at the point x. By the hypothesis of Problem A, $u(x, \varphi)$ is known for $x \in \partial D$, $\varphi \in (0, 2\pi)$, i.e., we have

$$u\big|_{\Gamma_1} = u_0(x, \varphi), \tag{1.0.3}$$

where $\Gamma_1 = \partial D \times (0, 2\pi)$. Thus, we come to the problem (1.0.2) – (1.0.3): Find the pair of functions (u, λ) from equation (1.0.2) that satisfy condition (1.0.3), where $f(x, \varphi)$ is a given sufficiently smooth function.

Now, suppose that we have the problem (1.0.2) – (1.0.3). Given the function f, we construct a set of curves $\{\Gamma\}$ such that f is the curvature of the curve that passes through $x \in D$ in the direction $(\sin\varphi, \cos\varphi)$. It is always possible to construct such a set of curves for a sufficiently smooth function $f(x, \varphi)$ with certain convexity properties. Integrating both sides of equality (1.0.2) along the curve $\gamma(x, \varphi)$ and observing (1.0.3), we arrive at Problem A. Thus we have proved that Problem A is equivalent to the problem (1.0.2) – (1.0.3) in the corresponding spaces. Reduction of a problem of integral geometry to a Dirichlet problem was first carried out in Lavrent'ev and Anikonov (1967).

The first two sections of this chapter are devoted to the analysis of the inverse problem for the transport equation, which is associated with a certain problem of integral geometry (Gel'fand, Graev, and Vilenkin, 1962; Lavrent'ev, Romanov, and Shishatskii, 1980). Since the generalization of the two-dimensional case to the three-dimentional case is nontrivial, we analyze them separately: the two-dimensional problem is treated in Section 1.1 and the three-dimensional one in Section 1.2.

Section 1.3 concerns two inverse problems for a special kinetic equation with close statements, and the problem of integral geometry along geodesics which is associated with these two problems.

Section 1.4 is devoted to a planar problem of integral geometry.

The existence and uniqueness theorems are proved using the general procedure described in Introduction. For each problem, we consruct an equation with properties 1 and 2 and the space in which it has a unique solution. Certain well known spaces that majorize the above-mentioned space from above and from below are pointed out. *A priori* estimates are used to prove the uniqueness theorems, and the Galerkin method is employed in the proofs of the existence theorems.

Throughout this chapter, D is a bounded simply connected domain.

1.1. TWO-DIMENSIONAL INVERSE PROBLEM FOR THE TRANSPORT EQUATION

1. Let $C^3_\pi(\Omega)$ denote the set of real-valued functions $u(x,\varphi)$ that are 2π-periodic in φ and three times continuously differentiable on Ω with respect to all arguments, where $\Omega = \{(x,\varphi) : x \in D \subset \mathbb{R}^2,\ \varphi \in (0, 2\pi),\ \partial D \in C^3\}$.

We introduce the following scalar product in $C^3_\pi(\Omega)$:

$$(u,z)_{1,c} = (u,z)_1 + \int_\Omega \left(\sum_{i=1}^{2} u_{x_i\varphi} z_{x_i\varphi} \right) \mathrm{d}\Omega,$$

where

$$(u,z)_1 = \int_\Omega \left(uz + \sum_{i=1}^{2} u_{x_i} z_{x_i} + u_\varphi z_\varphi \right) \mathrm{d}\Omega, \qquad \mathrm{d}\Omega = \mathrm{d}x\, \mathrm{d}\varphi.$$

Set $\|u\|_1 = [(u,u)_1]^{1/2}$ and $\|u\|_{1,c} = [(u,u)_{1,c}]^{1/2}$. Let $H^\pi_{1,c}(\Omega)$, $H^\pi_m(\Omega)$, and $H^\pi_{1,2}(\Omega)$ be the completions of $C^3_\pi(\Omega)$ with respect to the norms $\|\cdot\|_{1,c}$, $\|\cdot\|_{H^m}$ $(m = 1,2,3)$, and $\|\cdot\|_{1,2}$, respectively, where

$$\|\cdot\|^2_{1,2} = (u,u)_{1,2}, \qquad (u,u)_{1,2} = (u,u)_{1,c} + \int_\Omega u^2_{\varphi\varphi}\, \mathrm{d}\Omega.$$

Let $C^3_{\pi 0} = \{\psi : \psi|_{\partial D} = 0,\ \psi \in C^3_\pi\}$ and select a set $\{\omega_1, \omega_2, \omega_3, \dots\} \subset C^3_{\pi 0}$ which is complete and orthonormal in $L_2(\Omega)$. We may assume here that the linear span of this set is everywhere dense in $\overset{\circ}{H}{}^\pi_{1,2}(\Omega)$. Indeed, the space $H^\pi_{1,2} \cap \overset{\circ}{H}_1$ being separable, there exists a countable set $\{\varphi_i\}_{i=1}^\infty \subset C^3_{\pi 0}$ which is everywhere dense in this space. Extend this set up to a set which is everywhere dense in $L_2(\Omega)$, if necessary. Orthonormalizing the latter in $L_2(\Omega)$, we obtain $\{\omega_i\}_{i=1}^\infty$. Let $\mathcal{P}_n$ denote the orthogonal projector of $L_2(\Omega)$ onto $\mathfrak{M}_n$, where $\mathfrak{M}_n$ is the linear span of the vectors $\omega_1, \omega_2, \dots, \omega_n$. Furthermore, introduce the following designations:

$$Au \equiv \frac{\partial^2}{\partial l\, \partial \varphi}(Lu),$$

where

$$\frac{\partial}{\partial l} = \cos\varphi \frac{\partial}{\partial x_1} - \sin\varphi \frac{\partial}{\partial x_2},$$

$$Lu = u_{x_1} \sin\varphi + u_{x_2} \cos\varphi + u_\varphi f;$$

$\Gamma''(A)$ is the set of all functions $u(x,\varphi) \in L_2(\Omega)$ such that for any $u \in \Gamma''(A)$ there exists $y \in L_2(\Omega)$ with the following property:

$$\forall \eta \in C^3_{\pi 0}(\Omega) \quad \langle u, A^*\eta\rangle = \langle y, \eta\rangle,$$

where $\langle y, \eta\rangle$ is the scalar product in $L_2(\Omega)$, A^* is the differential expression conjugate to A in the sense of Lagrange; $y \overset{def}{=} Au$.

Suppose $\Gamma(A) \subset \Gamma''(A)$ and for any $u \in \Gamma(A)$ there exists a sequence $\{u_k\} \subset C^3_{\pi 0}$ such that $u_k \to u$ weakly in $L_2(\Omega)$ and $\langle Au_k, u_k\rangle \to \langle Au, u\rangle$ as $k \to \infty$. $\Gamma'(A)$ is the closure of $C^3_{\pi 0}$ with respect to the norm $\|u\|_{\Gamma(A)} = \|u\| + \|Au\|$ (here and in what follows, $\|\cdot\|$ denotes the norm in $L_2(\Omega)$). It is clear that $\Gamma'(A) \subset \Gamma(A) \subset \Gamma''(A)$. The following inclusion holds:

$$\Gamma''(A) \cap \overset{\circ}{H}{}^{\pi}_{1,c} \subset \Gamma(A) \subset L_2(\Omega),$$

where $\overset{\circ}{H}{}^{\pi}_{1,c}$ is the completion of $C^3_{\pi 0}$ with respect to the norm $\|\cdot\|_{1,c}$.

Evidently, only the leftmost part of the inclusion needs to be proved. Suppose $u \in \Gamma''(A) \cap \overset{\circ}{H}{}^{\pi}_{1,c}(\Omega)$. Then there exists a sequence $\{u_k\} \subset C^3_{\pi 0}$ such that $u_k \to u$ as $k \to \infty$ in the norm of $H^{\pi}_{1,c}$. Consequently, $u_k \to u$ as $k \to \infty$ in $L_2(\Omega)$ and

$$\langle Au_k, u_k\rangle = \big\langle Lu_k, \frac{\partial}{\partial\varphi}\frac{\partial}{\partial l}u_k\big\rangle \to \big\langle Lu, \frac{\partial}{\partial\varphi}\frac{\partial}{\partial l}u\big\rangle = \langle Au, u\rangle.$$

Hence $u \in \Gamma(A)$. The rightmost equality is true because $C^3_{\pi 0}$ is everywhere dense in $\overset{\circ}{H}{}^{\pi}_{1,c}$ and $u \in \Gamma''(A)$. Consider the following equation in Ω:

$$Lu \equiv u_{x_1}\sin\varphi + u_{x_2}\cos\varphi + u_\varphi f = \lambda(x,\varphi). \tag{1.1.1}$$

Its right-hand side is assumed to satisfy the equality

$$\langle \lambda, \frac{\partial}{\partial\varphi}\frac{\partial}{\partial l}\eta\rangle = 0 \tag{1.1.2}$$

for any function $\eta \in H^{\pi}_{1,2}(\Omega)$ such that $\eta = 0$ on $\Gamma_1 = \partial D \times (0, 2\pi)$, or, in other words, $\eta \in H^{\pi}_{1,2}(\Omega) \cap \overset{\circ}{H}_{1,\Gamma_1}(\Omega)$.

Problem 1.1.1. Find a pair of functions (u, λ) defined in Ω that satisfies equation (1.1.1), provided that the function $f \in C^2(\Omega)$ and the trace of the solution $u(x,\varphi)$ on Γ_1 $(u|_{\Gamma_1} = u_1)$ are known.

Remark 1.1.1. It is easy to verify that if $a(x_1, x_2)$ and $b(x_1, x_2)$ are continuous functions and $a(x)$ is continuously differentiable with respect to x_2, then the function

$$\lambda = \int_0^{\varphi} a(\varphi', x_1 \sin\varphi' + x_2 \cos\varphi')\, d\varphi' + b(x)$$

satisfies equality (1.1.2) for any $\eta \in H^{\pi}_{1,2}$ such that $\eta = 0$ on Γ_1.

It turned out that Problem 1.1.1 is associated with Problem A. The function f in Problem 1.1.1 and the set of curves $\{\Gamma\}$ in Problem A are interrelated (see Amirov 1984a, 1986a; Lavrent'ev and Anikonov 1967).

Consider the following problem together with Problem 1.1.1.

Problem 1.1.1′. Find a pair of functions (u, λ) satisfying the equation

$$Lu = \lambda + F, \tag{1.1.1'}$$

provided that the functions f and F are known, the trace of the solution of equation (1.1.1′) on Γ_1 is zero, and λ satisfies (1.1.2).

We shall see later that Problem 1.1.1 can be reduced to Problem 1.1.1′ if the boundary of D is of class C^3.

2. Now we turn to the uniqueness of solutions to Problem 1.1.1.

Theorem 1.1.1. *Let the following conditions hold:*

1. $f \in C^2(\overline{\Omega})$ *and*

$$\forall (x, \varphi) \in \Omega \quad \frac{\partial f}{\partial l} > f^2, \quad \text{where } \frac{\partial}{\partial l} = \cos\varphi \frac{\partial}{\partial x_1} - \sin\varphi \frac{\partial}{\partial x_2},$$

2. f *is* 2π*-periodic in* φ.

Then Problem 1.1.1 has at most one solution (u, λ) *such that* $u \in \Gamma(A)$ *and* $\lambda \in H^{-1}(\Omega)$.

Proof. It is sufficient to establish that the homogeneous version of Problem 1.1.1 has only trivial solution in $\Gamma(A) \times H^{-1}(\Omega)$. Let (u, λ) be a solution to Problem 1.1.1 such that $u \in \Gamma(A)$, $u_1 = 0$, and $\lambda \in H^{-1}(\Omega)$. From (1.1.1) and (1.1.2) it follows that $Au = 0$. Therefore, there exists a sequence $\{u_k\} \subset C^3_{0\pi}(\Omega)$ such that

$$\begin{aligned} &1.\ u_k \to u \quad \text{weakly in } L_2(\Omega) \\ &2.\ \langle Au_k, u_k \rangle \to 0 \end{aligned} \tag{1.1.3}$$

The following equality can be easily verified:

$$-\langle Au_k, u_k\rangle \tag{1.1.4}$$
$$= \frac{1}{2}\int_\Omega \Big[|\nabla_x u_k|^2 + u_{k\varphi}^2 \frac{\partial f}{\partial l} + 2f u_{k\varphi}(u_{kx_1}\sin\varphi + u_{kx_2}\cos\varphi)\Big]\, d\Omega.$$

Indeed, we have

$$-Au_k u_k = \frac{\partial}{\partial\varphi}(Lu_k)\frac{\partial}{\partial l}u_k - \frac{\partial}{\partial l}\left[\frac{\partial}{\partial\varphi}(Lu_k)u_k\right]. \tag{1.1.5}$$

On the other hand, the following identity holds:

$$2\frac{\partial}{\partial\varphi}(Lu_k)\frac{\partial}{\partial l}u_k$$
$$= |\nabla_x u_k|^2 + u_{k\varphi}^2\frac{\partial f}{\partial l} + 2f u_{k\varphi}(u_{kx_1}\sin\varphi + u_{kx_2}\cos\varphi)$$
$$- \frac{\partial}{\partial x_1}(u_{kx_2}u_{k\varphi} + u_{k\varphi}^2 f\cos\varphi) + \frac{\partial}{\partial x_2}(u_{kx_1}u_{k\varphi} + f u_{k\varphi}^2\sin\varphi)$$
$$+ \frac{\partial}{\partial\varphi}(u_{kx_1}^2\sin\varphi\cos\varphi + u_{kx_1}u_{kx_2}\cos 2\varphi + 2u_{kx_1}u_{k\varphi}f\cos\varphi$$
$$- u_{kx_2}^2\sin\varphi\cos\varphi - 2u_{kx_2}u_{k\varphi}f\sin\varphi). \tag{1.1.6}$$

Then, taking into account that $u_k = 0$ on Γ_1 and u_k is 2π-periodic in φ, we see that (1.1.5) and (1.1.6) imply (1.1.4).

It is easily verified that the quadratic form $J(\nabla u_k)$ in u_{kx_1}, u_{kx_2}, $u_{k\varphi}$, defined by the equality

$$J(\nabla u) = |\nabla_x u|^2 + u_\varphi^2\frac{\partial f}{\partial l} + 2f u_\varphi(u_{x_1}\sin\varphi + u_{x_2}\cos\varphi),$$

is positive definite if $\partial f/\partial l > f^2$. Since the domain D is bounded and $u_k = 0$ on Γ_1, we can easily prove that $\|u_k\| \le C\int J(\nabla u_k)\, d\Omega$, where $C > 0$ depends on the size of D and does not depend on k. Consequently, by virtue of (1.1.3) and (1.1.4) we have

$$\|u\|^2 \le \underline{\lim}_{k\to\infty}\|u_k\|^2 \le C\,\underline{\lim}_{k\to\infty}\int_\Omega J(\nabla u_k)\, d\Omega = -C\lim_{k\to\infty}\langle Au_k, u_k\rangle = 0 \tag{1.1.7}$$

Inequality (1.1.7) implies $\|u\| = 0$, i.e., $u = 0$, and equation (1.1.1) implies $\lambda = 0$. This completes the proof of the theorem. □

3. In this part of the section we establish the existence theorem for Problem 1.1.1′.

Theorem 1.1.2. *Under the assumptions of Theorem 1.1.1, suppose that* $F \in H_2^{\pi}(\Omega)$. *Then there exists a solution* (u, λ) *of Problem* 1.1.1′ *that satisfies the following conditions:*

$$\begin{gathered} u \in \Gamma(A), \quad u \in H_1^{\pi}(\Omega), \quad \lambda \in L_2(\Omega), \\ \|u\|_{H_1^{\pi}(\Omega)} + \|\lambda\| \leq C(\|F_{\varphi}\| + \|F\|), \end{gathered} \tag{1.1.8}$$

where $C > 0$ *depends on* f *and the size of* D.

Proof. Consider the following problem: Solve the equation

$$Au = \mathcal{F}, \tag{1.1.9}$$

with the condition

$$u\big|_{\Gamma_1} = 0, \tag{1.1.10}$$

where

$$Au = \frac{\partial}{\partial l}\frac{\partial}{\partial \varphi}(Lu), \quad \mathcal{F} = \frac{\partial}{\partial l}\frac{\partial}{\partial \varphi}F.$$

We shall call it the problem (1.1.9) – (1.1.10) for brevity.

An approximate solution of order N of the problem (1.1.9) – (1.1.10),

$$u_N = \sum_{i=1}^{N} \alpha_{N_i}\omega_i(x, \varphi); \quad \alpha_N = (\alpha_{N_1}, \alpha_{N_2}, \dots, \alpha_{N_N}),$$

is defined as a solution to the following problem:

Find the vector α_N from the system of linear algebraic equations

$$\langle Au_N, \omega_j\rangle = \langle \mathcal{F}, \omega_j\rangle, \quad j = 1, \dots, N \tag{1.1.11}$$

or

$$\sum_{j=1}^{N} \alpha_{N_j}\langle A\omega_j, \omega_i\rangle = \langle \mathcal{F}, \omega_i\rangle, \quad i = 1, \dots, N \tag{1.1.12}$$

We shall prove that there exists a unique solution α_N of the system (1.1.12) for any $F \in H_2^{\pi}(\Omega)$ under the hypotheses of Theorem 1.1.2. To this end, it suffices to prove that the homogeneous version of system (1.1.12) has only trivial solution. Assume the contrary. Let the homogeneous version of system (1.1.12) have a nonzero solution $\bar{\alpha}_N = (\bar{\alpha}_{N_1}, \bar{\alpha}_{N_2}, \dots, \bar{\alpha}_{N_N})$. In the system (1.1.12) with $F = 0$, substituting $\bar{\alpha}_N$ for α_N, multiplying the ith equation by $-2\bar{\alpha}_{N_i}$ and summing with respect to i from 1 to N, we obtain

$$-2\langle A\bar{u}_N, \bar{u}_N\rangle = 0, \tag{1.1.13}$$

where $\bar{u}_N = \sum_{i=1}^N \bar{\alpha}_{N_i}\omega_i$. By (1.1.4),

$$-2\langle A\bar{u}_N, \bar{u}_N\rangle = \int_\Omega J(\nabla \bar{u}_N)\, d\Omega.$$

Therefore, since the quadratic form $J(\nabla\bar{u}_N)$ is positive definite and $\bar{u}_N = 0$ on Γ_1, we have $\bar{u}_N = 0$ in Ω, and $\bar{\alpha}_{N_i} = 0$ because the system $\{\omega_i\}$ is linearly independent. This contradicts the condition $\bar{\alpha}_N \neq 0$.

Thus, system (1.1.12) has a unique solution α_N for any $F \in H_2^\pi(\Omega)$.

Since $f \in C^2(\Omega)$ and $\partial f/\partial l > f^2$, there exists a number $\mu > 0$ such that $J(\nabla u) \geq \mu\|\nabla u\|^2$ for any $u \in C^1(\Omega)$, where $\nabla u = (u_{x_1}, u_{x_2}, u_\varphi)$.

Now we estimate the solution u_N of system (1.1.11) in terms of $\mathcal{F}$. For this purpose, we multiply the ith equation of system (1.1.11) by $-2\alpha_{N_i}$ and sum the obtained relations with respect to i from 1 to N. Then we have

$$\left\langle \frac{\partial}{\partial l}\frac{\partial}{\partial \varphi}(Lu_N), u_N\right\rangle = -2\langle \mathcal{F}, u_N\rangle. \tag{1.1.14}$$

Since $\mathcal{F} = \dfrac{\partial}{\partial l}\dfrac{\partial}{\partial \varphi}F$, the right-hand side of (1.1.14), by virtue of the condition that $u_N = 0$ on Γ_1, is estimated as follows:

$$\begin{aligned}\left|-2\int_\Omega \mathcal{F}u_N\, dx\, d\varphi\right| &= 2\left|\int_\Omega \mathcal{F}_\varphi \frac{\partial}{\partial l}u_N\, dx\, d\varphi\right| \\ &\leq \beta\int_\Omega \mathcal{F}_\varphi^2\, dx\, d\varphi + \beta^{-1}\int_\Omega |\nabla_x u_N|^2\, dx\, d\varphi.\end{aligned}$$

Here $\beta^{-1} < \mu$. For $u_N = \bar{u}_N$, as was shown above, the left-hand side of (1.1.14) equals $\int_\Omega J(\nabla\bar{u}_N)\, d\Omega$. Therefore, from (1.1.14) we get

$$\int_\Omega J(\nabla u_N)\, d\Omega \leq \beta\int_\Omega F_\varphi^2\, d\Omega + \beta^{-1}\int_\Omega |\nabla_x u_N|^2\, d\Omega. \tag{1.1.15}$$

Since Ω is bounded and $u_N = 0$ on Γ_1, the foregoing inequality implies

$$\|u_N\|_{\overset{\circ}{H}{}_1^\pi(\Omega)} \leq C\,\|F_\varphi\|, \tag{1.1.16}$$

where the constant $C > 0$ is independent of N.

Thus, the set of functions u_N, $N = 1, 2, 3, \ldots$, is bounded in $\overset{\circ}{H}{}_1^\pi\ (\Omega)$. Because $\overset{\circ}{H}{}_1^\pi\ (\Omega)$ is a Hilbert space, $\{u_N\}$ is weakly compact in it. Hence, there exists a subsequence in this set, which we again denote by $\{u_N\}$, that converges weakly in $\overset{\circ}{H}{}_1^\pi\ (\Omega)$ to a certain function $u \in \overset{\circ}{H}{}_1^\pi\ (\Omega)$.

From inequality (1.1.16) and weak convergence of $\{u_k\}$ to u in $\overset{\circ}{H}{}_1^\pi\ (\Omega)$ it follows that

$$\|u\|_{\overset{\circ}{H}{}_1^\pi(\Omega)} \leq \varliminf_{N\to\infty} \|u_N\|_{\overset{\circ}{H}{}_1^\pi(\Omega)} \leq C\,\|F_\varphi\|.$$

Since $u \in \overset{\circ}{H}{}_1^\pi\ (\Omega)$, by the definition of $\overset{\circ}{H}{}_1^\pi\ (\Omega)$ we have $u = 0$ on Γ_1. Using estimate (1.1.16), it is easy to prove that there exists a subsequence of $\{u_N\}$, which we again denote by u_N, such that u_{Nx_i} and $u_{N\varphi}$, $i = 1, 2$, weakly converge in $L^2(\Omega)$ to u_{x_i} and u_φ, respectively, as $N \to \infty$. Transferring the appropriate derivatives to ω_i in (1.1.11) and observing that $u_N = 0$ on Γ_1 and the functions u_N and F are 2π-periodic together with their derivatives up to the second order, we have

$$\left\langle Lu_N - F, \frac{\partial}{\partial\varphi}\frac{\partial}{\partial l}(\omega_i) \right\rangle = 0 \quad \text{for } N \geq i.$$

Since the linear span of the functions ω_i, $i = 1, 2, 3, \ldots$, is everywhere dense in $\overset{\circ}{H}{}_{1,2}^\pi\ (\Omega)$, passing to the limit as $N \to \infty$ in the foregoing equality, we obtain

$$\left\langle Lu - F, \frac{\partial}{\partial\varphi}\frac{\partial}{\partial l}\eta \right\rangle = 0 \tag{1.1.17}$$

for any $\eta \in \overset{\circ}{H}{}_{1,2}^\pi\ (\Omega)$. Setting $\lambda = Lu - F$, from (1.1.17) we see that λ satisfies the condition (1.1.12) for any $\eta \in \overset{\circ}{H}{}_{1,2}^\pi\ (\Omega)$, and the following estimate is valid:

$$\|\lambda\| \leq C\|u\|_{H_1^\pi(\Omega)} + \|F\|.$$

Consequently, using the inequality $\|u\|_{H_1^\pi(\Omega)} \leq C\,\|F_\varphi\|$, we obtain (1.1.8). In the expressions above, C stands for different constants that depend only on the given functions and the size of the domain D.

Thus, we have found a solution (u, λ) of Problem 1.1.1′, where $u \in \overset{\circ}{H}{}_1^\pi(\Omega)$ and $\lambda \in L_2(\Omega)$.

Now we prove that $u \in \Gamma(A)$. Since $u \in L_2(\Omega)$ and $F \in H_2^\pi(\Omega)$, from (1.1.17) it follows that $\mathcal{F} = Au \in L_2(\Omega)$ in the generalized sense, i.e., $u \in \Gamma''(A)$. Indeed, for any $\eta \in C_0^\infty \subset \overset{\circ}{H}{}_{1,2}^\pi(\Omega)$ we have

$$\langle u, A^*\eta\rangle = \Big\langle u, L^*\frac{\partial}{\partial\varphi}\frac{\partial}{\partial l}\eta\Big\rangle = \Big\langle F, \frac{\partial}{\partial\varphi}\frac{\partial}{\partial l}\eta\Big\rangle = \langle \mathcal{F}, \eta\rangle,$$

where

$$\mathcal{F} = \frac{\partial}{\partial\varphi}\frac{\partial}{\partial l}F \in L_2(\Omega).$$

To complete the proof of Theorem 1.1.2, it remains to prove the convergence $\langle Au_k, u_k\rangle \to \langle Au, u\rangle$ as $N \to \infty$. From (1.1.11) it follows that $\mathcal{P}_N Au_N = \mathcal{P}_N\mathcal{F}$. Since $\mathcal{P}_N$ is an orthogonal projector, $\mathcal{P}_N\mathcal{F}$ strongly converges to $\mathcal{F}$ in $L_2(\Omega)$ as $N \to \infty$, i.e., $\mathcal{P}_N Au_N \to \mathcal{F} = Au$ strongly in $L_2(\Omega)$ as $N \to \infty$. Then we have $\langle \mathcal{P}_N Au_N, u_N\rangle \to \langle Au, u\rangle$ as $N \to \infty$ because $\{u_N\}$ weakly converges to u and $\{\mathcal{P}_N Au_N\}$ strongly converges to Au in $L_2(\Omega)$ as $N \to \infty$. By the definition of $\mathcal{P}_N$ and u_N (since the operator $\mathcal{P}_N$ is self-adjoint in L_2),

$$\langle Au_N, u_N\rangle = \langle Au_N, \mathcal{P}_N u_N\rangle = \langle \mathcal{P}_N Au_N, u_N\rangle.$$

Hence $\langle Au_N, u_N\rangle \to \langle Au, u\rangle$, which completes the proof. □

Remark 1.1.2. It is easy to see from the proofs of Theorems 1.1.1 and 1.1.2 that the same results can be obtained for a similar problem if we impose more restrictions on f and replace equation (1.1.1) with the general kinetic equation, which involves the terms associated with absorption and scattering, i.e.,

$$Lu + \beta u + \int_0^{2\pi} K(x, \varphi, \varphi')u(x, \varphi')\, d\varphi' = \lambda(x, \varphi),$$

where $\beta(x, \varphi)$ and $K(x, \varphi, \varphi')$ are given functions which are continuously differentiable with respect to all their arguments.

Consider the question of reducing Problem 1.1.1 to Problem 1.1.1′. Suppose that $\partial D \in C^3$ and let $C_\pi^3(\Gamma_1)$ be the set of traces of functions from $C_\pi^3(\Omega)$ on Γ_1. By $H_3^\pi(\Gamma_1)$ we denote the closure of $C_\pi^3(\Gamma_1)$ with respect to the norm of the space $H_3(\Gamma_1)$. Obviously, $H_3^\pi(\Gamma_1)$ is a Hilbert functional space.

For $u_1 \in H_3^\pi(\Gamma_1)$, there exists a function $w(x,\varphi) \in H_3^\pi(\Omega)$ such that $w = u_1$ on Γ_1. Indeed, for any fixed "section" $\widetilde{\Gamma}_1 = \partial D \times \varphi_0$ of the surface Γ_1, it is possible to construct a function $w(x,\varphi_0)$ (in the same way as in Theorem 2, Mikhailov, 1976, p. 131) such that $w(x,\varphi_0) \in H_3(D)$, $w(x,\varphi_0) = u_1(x,\varphi_0)$ on $\widetilde{\Gamma}_1$, and

$$\|w(x,\varphi_0)\|_{H_3(\Gamma_1)} \leq C\, \|u_1(x,\varphi_0)\|_{H_3(\widetilde{\Gamma}_1)}.$$

Thus, since $u_1 \in H_3^\pi(\Gamma_1)$, from the construction procedure it is clear that $w(x,\varphi)$ belongs to the space $H_3^\pi(\Omega)$ and, moreover,

$$\|w\|_{H_3^\pi(\Omega)} \leq C\, \|u_1\|_{H_3(\Gamma_1)}.$$

Remark 1.1.3. It is easy to see from the proofs of Theorems 1.1.1 and 1.1.2 that they are also valid in the case $f = 0$. Then $J(\nabla u) = |\nabla_x u|^2$ and the solution $u(x,\varphi)$ of Problem 1.1.1 belongs to $H_{1,x}^\pi(\Omega)$, which is the space of functions such that their generalized derivatives with respect to x_1 and x_2 belong to $L_2(\Omega)$. Moreover, the term u_φ^2 will not be present in the expression for $\|u\|_{H_1^\pi(\Omega)}$ obtained from (1.1.8).

1.2. THREE-DIMENSIONAL INVERSE PROBLEM FOR THE TRANSPORT EQUATION

In the second part of this section we analyze the statement of the three-dimensional inverse problem for the transport equation, its relation to the problem of integral geometry, and an identity associated with the three-dimemsional transport equation.

1. Let S^2 be the unit sphere and D be a domain in $\mathbb{R}^3$ with sufficiently smooth boundary. Set

$$\Omega = \{(x,\nu) \,:\, x \in D, \nu \in S^2\},$$

$$\nu = (\cos\varphi_1, \cos\varphi_2 \sin\varphi_1, \sin\varphi_2 \sin\varphi_1),$$

$$\frac{\partial}{\partial\theta_1} = \frac{\partial}{\partial\varphi_1}, \quad \frac{\partial}{\partial\theta_2} = \frac{1}{\sin\varphi_1}\frac{\partial}{\partial\varphi_2}, \quad \varphi_1 \in (0,\pi),\ \varphi_2 \in (0,2\pi),$$

$$\Gamma = \partial D \times (0,\pi) \times (0,2\pi), \quad \varphi = (\varphi_1,\varphi_2).$$

By $\mathcal{M}$ we denote the set of all finite linear combinations of the functions

$$\omega(x)P_{n,m}(\cos\varphi_1)\sin m\varphi_2, \quad \omega(x)P_{n,m}(\cos\varphi_1)\cos m\varphi_2,$$

where $n, m = 0, 1, 2, \dots$, $P_{n,0}(x) \equiv P_n(x)$, $P_n(x)$ are Legendre polynomials,

$$P_{n,m}(x) = (1 - x^2)^{m/2} \frac{\mathrm{d}^m}{\mathrm{d}x^m} P_n(x),$$

and $\omega(x)$ is an arbitrary real-valued function of class $C^3(D)$. $\overline{H}_1(\Omega)$ is the closure of $\mathcal{M}$ with respect to the norm $\|u\|_{\overline{1}} = [(u,u)_{\overline{1}}]^{1/2}$, where

$$(u,z)_{\overline{1}} = \int_\Omega \Big(uz + \sum_{i=1}^{3} u_{x_i} z_{x_i} + \sum_{i=1}^{2} u_{\theta_i} z_{\theta_i} \Big)\, \mathrm{d}\Omega,$$

$\mathrm{d}x$ is the volume element of the domain D, $\mathrm{d}S^2$ is the surface element of the sphere S^2.

$\overline{H}_{1,2}(\Omega)$ is a real Hilbert space, the completion of the set $\mathcal{M}$ with respect to the norm $\|u\|_{\overline{1,2}} = [(u,u)_{\overline{1,2}}]^{1/2}$, where $(u,z)_{\overline{1,2}} = (u,z)_{\overline{1}} + (u,z)_{\widetilde{1,2}}$,

$$(u,z)_{\widetilde{1,2}} = \int_\Omega \Big(\sum_{i=1}^{3} \sum_{j=1}^{2} z_{x_i \theta_j} u_{x_i \theta_j} \Big)\, \mathrm{d}\Omega,$$

$$\widetilde{C}_0^2(D) = \{ w(x) \,:\, w \in C^2(D), w = 0 \text{ on } \partial D \},$$

$$\mathcal{M}^0 = \{ \eta \,:\, \eta \in \mathcal{M}, \eta = 0 \text{ on } \Gamma \},$$

$\overline{H}_1^0(\Omega)$ is the completion of $\mathcal{M}^0$ with respect to the norm $\|\cdot\|_{\overline{1,2}}$.

We also introduce the following designations:

$$\nu' = (-\sin\varphi_1,\ \cos\varphi_1\cos\varphi_2,\ \cos\varphi_1\sin\varphi_2), \quad \nu'' = (0,\ -\sin\varphi_2,\ \cos\varphi_2),$$

$$f = (f_1,\ f_2,\ f_3), \quad F_1 = f\cdot\nu', \quad F_2 = f\cdot\nu''.$$

Consider the equation

$$L_1 u \equiv \nu\cdot\nabla_x u + u_{\theta_1} F_1 + u_{\theta_2} F_2 = \lambda \tag{1.2.1}$$

in the domain Ω, where $\nu\cdot\nabla_x u$ is the scalar product of the vectors ν and $\nabla_x u$ in $\mathbb{R}^3$ and the right-hand side λ is such that

$$\int_\Omega \lambda \Big[\frac{\partial}{\partial\theta_1}(\nabla_x\eta\cdot\nu') + (\nabla_x\eta\cdot\nu')\cot\varphi_1 + \frac{\partial}{\partial\theta_2}(\nabla_x\eta\cdot\nu'') \Big]\, \mathrm{d}\Omega = 0 \tag{1.2.2}$$

for any function $\eta \in \overline{H}_{1,2}(\Omega)$ vanishing on Γ.

We set

$$Au \equiv \left(-\sin\varphi_2 \frac{\partial}{\partial x_2} + \cos\varphi_2 \frac{\partial}{\partial x_3}\right) \frac{\partial}{\partial\theta_2}(L_1 u)$$

$$+ \left(-\sin\varphi_1 \frac{\partial}{\partial x_1} + \cos\varphi_1 \cos\varphi_2 \frac{\partial}{\partial x_2}\right.$$

$$\left. + \cos\varphi_1 \sin\varphi_2 \frac{\partial}{\partial x_3}\right) \frac{\partial}{\partial\theta_1}(L_1 u). \qquad (*)$$

Let $\Gamma''(A)$ be a set of functions with the following property: for any $u \in \Gamma''(A)$, $u \in \overline{H}_1^0(\Omega)$ and there exists a function $y \in L_2(\Omega)$ such that $\forall \eta \in \mathcal{M}^0$

$$\left\langle L_1 u, \left[\frac{\partial}{\partial\theta_1}(\nabla_x \eta \cdot \nu') + (\nabla_x \eta \cdot \nu') \cot\varphi_1 + \frac{\partial}{\partial\theta_2}(\nabla_x \eta \cdot \nu'')\right]\right\rangle = \langle y, \eta\rangle$$

Set $y = Au$. This assumption becomes clear from the observation that $\mathrm{d}\Omega = \sin\varphi_1 \, \mathrm{d}x \, \mathrm{d}\varphi_1 \, \mathrm{d}\varphi_2$ and for u, $\eta \in \mathcal{M}^0$

$$L_1 u \frac{\partial}{\partial\theta_1}(\nabla_x \eta \cdot \nu') + L_1 u \, (\nabla_x \eta \cdot \nu') \cot\varphi_1$$

$$= \frac{1}{\sin\varphi_1}\left[L_1 u \frac{\partial}{\partial\varphi_1}(\nabla_x \eta \cdot \nu') \sin\varphi_1 + L_1 u \, (\nabla_x \eta \cdot \nu) \cos\varphi_1\right]$$

$$\equiv \frac{1}{\sin\varphi_1}\left\{\frac{\partial}{\partial\varphi_1}\left[L_1 u \, (\nabla_x \eta \cdot \nu') \sin\varphi_1\right]\right\} - \frac{\partial}{\partial\varphi_1}(L_1 u)(\nabla_x \eta \cdot \nu').$$

The definition of the set $\Gamma''(A)$ given above is slightly different from that given in Section 1.1. It is caused by the terms like $\partial(\cot\varphi_1)/\partial\theta_1$ appearing in the expression for the conjugate operator A^* in the space $L_2(\Omega)$ due to the relation $\mathrm{d}\Omega = \sin\varphi_1 \, \mathrm{d}x \, \mathrm{d}\varphi_1 \, \mathrm{d}\varphi_2$. The definition of $\Gamma(A)$ and $\Gamma'(A)$ is the same as in Section 1.1 with $\mathcal{M}^0$ substituted for $C_{\pi 0}^3$ and the expression given by the formula $(*)$ substituted for the differential expression A. It is obvious that $\overline{H}_1^0(\Omega) \cap \Gamma'(A) \subset \Gamma(A) \subset \Gamma''(A)$. Arguments similar to those used in Section 1.1 can be used to prove that $\Gamma''(A) \cap \overline{H}_{1,2}^0(\Omega) \subset \Gamma(A) \subset L_2(\Omega)$.

Choose a set $\{\omega_1, \omega_2, \omega_3, \dots\} \subset \widetilde{C}_0^2(D)$ orthonormal in $L_2(\Omega)$. It is known (see Smirnov, 1969) that the system of functions

$$\{P_n(\cos\varphi_1),\ P_{n,m}(\cos\varphi_1)\sin m\varphi_2,\ P_{n,m}(\cos\varphi_1)\cos m\varphi_2\},$$

where $n = 1, 2, \dots$, $m = 1, 2, \dots, n$, is a complete orthogonal system in $L_2(S^2)$. Assume that the linear span of the set $\{\omega_1, \omega_2, \omega_3, \dots\}$ is everywhere dense in $\overset{\circ}{H}_{1,2}(D)$. Let $\mathcal{P}_N$ denote the orthogonal projector of the space $L_2(\Omega)$ onto $\mathfrak{M}_N$, where $\mathfrak{M}_N$ is the linear span of the system of vectors $\omega_i P_n$, $\omega_i \cos m\varphi_2 P_{n,m}(\cos\varphi_1)$, $\omega_i \sin m\varphi_2 P_{n,m}(\cos\varphi_1)$, $i, n = 1, 2, 3, \dots, N$, $m = 1, 2, 3, \dots, n$.

Problem 1.2.1. Determine the pair of functions (u, λ) defined in Ω that satisfies equation (1.2.1) from the given functions F_1, F_2, and the trace of $u(x, \nu)$ on Γ $(u\big|_\Gamma = u_0)$.

Remark 1.2.1. The function λ of the form $\lambda = g(x) + \psi(\varphi)$, where $g, \psi \in C^1$, satisfies relation (1.2.2) for any function $\eta \in \overline{H}_{1,2}(\Omega)$ such that $\eta = 0$ on Γ.

As in the planar case, Problem 1.2.1 is associated with a problem of integral geometry analogous to Problem $\mathcal{A}$. We shall call it Problem $\mathcal{B}$. The latter is formulated similarly to Problem $\mathcal{A}$, but the dimension of the space and the way of specifying the set of curves $\{\Gamma\}$ are different. In this case, the curves are specified using the angle variables $\varphi = (\varphi_1, \varphi_2)$, and these angles are defined as the solution to the Cauchy problem

$$\frac{\partial\tilde{\varphi}}{\partial s} = F(x, \tilde{\varphi}), \qquad \tilde{\varphi}(0) = \varphi,$$

where $\tilde{\varphi} = (\varphi_1, \varphi_2)$ and $F = (F_1, F_2)$.

The following lemma can be easily verified.

Lemma 1.2.1. *Let $u(x, \nu)$ be twice continuously differentiable in Ω. Then the following identity holds:*

$$L_2 L_1 u = J(u) + \sum_{i=1}^{5} \mathrm{d}_i(u),$$

with $J(u)$, L_1, L_2, and d_i defined as follows:

$$\frac{1}{2} L_2 \equiv (\nabla_x u \cdot \nu') \frac{\partial}{\partial\theta_1} + (\nabla_x u \cdot \nu'') \frac{\partial}{\partial\theta_2},$$

$$
\begin{aligned}
J(u) &= |\nabla_x u|^2 + (u_{x_1} \cos\varphi_1 + u_{x_2} \sin\varphi_1 \cos\varphi_2 + u_{x_3} \sin\varphi_1 \sin\varphi_2)^2 \\
&+ u_{\varphi_1}^2 (f_{1x_1} \sin^2\varphi_1 - f_{2x_1} \cos\varphi_1 \sin\varphi_1 \cos\varphi_2 - f_{3x_1} \sin\varphi_1 \sin\varphi_2 \cos\varphi_1 \\
&\quad - f_{1x_2} \sin\varphi_1 \cos\varphi_1 \cos\varphi_2 + f_{3x_2} \cos^2\varphi_1 \sin\varphi_2 \cos\varphi_2 \\
&\quad + f_{2x_2} \cos^2\varphi_1 \cos^2\varphi_2 - f_{1x_3} \sin\varphi_1 \cos\varphi_1 \sin\varphi_2 \\
&\quad + f_{2x_3} \cos^2\varphi_1 \sin\varphi_2 \cos\varphi_2 + f_{3x_3} \cos^2\varphi_1 \sin^2\varphi_2) \\
&+ u_{\theta_2}^2 (f_{2x_2} \sin^2\varphi_2 - f_{3x_2} \cos\varphi_2 \sin\varphi_2 + f_{3x_3} \cos^2\varphi_2 - f_{2x_3} \cos\varphi_2 \sin\varphi_2) \\
&+ u_{x_1} u_{\varphi_1} (-2f_1 \sin 2\varphi_1 + 4f_2 \cos^2\varphi_1 \cos\varphi_2 - f_2 \cos\varphi_2 \\
&\quad + 4f_3 \cos^2\varphi_1 \sin\varphi_2 - f_3 \sin\varphi_2 + f_{3\varphi_2} \cos\varphi_2 - f_{2\varphi_2} \sin\varphi_2) \\
&+ u_{x_1} u_{\varphi_2} (-3f_2 \cot\varphi_1 \sin\varphi_2 + 3f_3 \cot\varphi_1 \cos\varphi_2 \\
&\quad + f_{2\varphi_1} \sin\varphi_2 - f_{3\varphi_1} \cos\varphi_2) \\
&+ u_{x_2} u_{\varphi_1} (2f_1 \cos 2\varphi_2 \cos\varphi_2 - f_1 \cos\varphi_2 + 2f_2 \sin 2\varphi_1 \cos^2\varphi_2 \\
&\quad + f_3 \sin 2\varphi_1 \cos 2\varphi_1 + f_{1\varphi_2} \sin\varphi_2 - f_{3\varphi_2} \cot\varphi_1) \\
&+ u_{x_2} u_{\varphi_2} (f_3 \cos 2\varphi_2 + f_1 \cot\varphi_1 \sin\varphi_2 - 2f_2 \sin 2\varphi_2 \\
&\quad - f_{1\varphi_1} \sin\varphi_2 + f_{3\varphi_1} \cot\varphi_1) \\
&+ u_{x_3} u_{\varphi_1} (2f_1 \cos 2\varphi_1 \sin\varphi_2 - f_1 \sin\varphi_2 + f_2 \sin 2\varphi_1 \sin 2\varphi_2 \\
&\quad + 2f_3 \sin 2\varphi_1 \sin^2\varphi_2 + f_{2\varphi_2} \cot\varphi_1 - f_{1\varphi_2} \cos\varphi_2) \\
&+ u_{x_3} u_{\varphi_2} (f_{1\varphi} \cos\varphi_2 + f_1 \cot\varphi_1 \cos\varphi_2 + f_2 \cos 2\varphi_2 \\
&\quad + 2f_3 \sin 2\varphi_2 - f_{2\varphi_1} \cot\varphi_1) \\
&+ u_{\varphi_1} u_{\varphi_2} (f_{2x_1} \sin\varphi_2 - f_{3x_1} \cos\varphi_2 + f_{1x_2} \sin\varphi_2 \\
&\quad - f_{2x_2} \cot\varphi_1 \sin 2\varphi_2 + f_{3x_2} \cot\varphi_1 \cos 2\varphi_2 - f_{1x_3} \cos\varphi_2 \\
&\quad + f_{2x_3} \cot\varphi_1 \cos 2\varphi_2 + f_{3x_3} \cot\varphi_1 \sin 2\varphi_2),
\end{aligned}
$$

$$
\begin{aligned}
\mathrm{d}_1(u) = \frac{\partial}{\partial x_1} \Big[& u_{x_2} u_{\varphi_1} \cos\varphi_2 + u_{x_3} u_{\varphi_1} \sin\varphi_2 \\
& - u_{x_2} u_{\varphi_2} \cot\varphi_1 \sin\varphi_2 + u_{x_3} u_{\varphi_2} \cot\varphi_1 \cos\varphi_2 \\
& + u_{\theta_1}^2 (-f_1 \sin^2\varphi_1 + f_2 \sin\varphi_1 \cos\varphi_1 \cos\varphi_2 \\
& + f_3 \sin\varphi_1 \cos\varphi_1 \sin\varphi_2) + u_{\varphi_1} u_{\varphi_2} (f_3 \cos\varphi_2 - f_2 \sin\varphi_2) \Big],
\end{aligned}
$$

$$\begin{aligned}
\mathrm{d}_2(u) = \frac{\partial}{\partial x_2}\Big[& -u_{x_1}u_{\varphi_1}\cos\varphi_2 + u_{x_1}u_{\varphi_2}\frac{\cos\varphi_2}{\sin\varphi_1}\sin\varphi_2 \\
& + u_{x_3}u_{\varphi_2} + u^2_{\varphi_1}(f_1\sin\varphi_1\cos\varphi_1\cos\varphi_2 \\
& - f_2\cos^2\varphi_2\cos^2\varphi_1 - f_3\cos^2\varphi_1\sin\varphi_2\cos\varphi_2) \\
& + u_{\varphi_1}u_{\varphi_2}(f_2\cot\varphi_1\sin 2\varphi_2 - f_1\sin\varphi_2 - f_3\cot\varphi_1\cos^2\varphi_2 \\
& + f_3\cot\varphi_1\sin^2\varphi_2) + u^2_{\theta_2}(-f_2\sin^2\varphi_2 + f_3\cos\varphi_2\sin\varphi_2)\Big],
\end{aligned}$$

$$\begin{aligned}
\mathrm{d}_3(u) = \frac{\partial}{\partial x_3}\Big[& -u_{x_1}u_{\varphi_1}\sin\varphi_2 - u_{x_1}u_{\theta_2}\cos\varphi_1\cos\varphi_2 - u_{x_2}u_{\varphi_2} \\
& + u^2_{\theta_1}(-f_3\cos^2\varphi_1\sin^2\varphi_2 + f_1\sin\varphi_1\cos\varphi_1\sin\varphi_2 \\
& - f_2\cos^2\varphi_1\sin\varphi_2\cos\varphi_2) + u_{\varphi_1}u_{\varphi_2}(f_1\cos\varphi_2 \\
& + f_2\cot\varphi_1\sin^2\varphi_2 - f_2\cot\varphi_1\cos^2\varphi_2 - 2f_3\cot\varphi_1\sin\varphi_2\cos\varphi_2) \\
& + u^2_{\theta_2}(f_2\sin\varphi_2\cos\varphi_2 - f_3\cos^2\varphi_2)\Big],
\end{aligned}$$

$$\begin{aligned}
\mathrm{d}_4(u) = \frac{1}{\sin\varphi_1}\frac{\partial}{\partial\varphi_1}\Big[& -u^2_{x_1}\cos\varphi_1\sin^2\varphi_1 \\
& + u_{x_1}u_{x_2}\sin\varphi_1\cos\varphi_2\cos 2\varphi_1 + u_{x_1}u_{x_3}\sin\varphi_1\cos 2\varphi_1\sin\varphi_2 \\
& + u^2_{x_2}\sin^2\varphi_1\cos\varphi_1\cos^2\varphi_2 + u_{x_1}u_{x_3}\sin^2\varphi_1\cos\varphi_1\sin 2\varphi_2 \\
& + u^2_{x_3}\sin^2\varphi_1\cos\varphi_1\sin^2\varphi_2 + u_{x_1}u_{\varphi_1}(2f_1\sin^3\varphi_1 \\
& - 2f_2\cos\varphi_1\sin^2\varphi_1\cos\varphi_2 - 2f_3\sin^2\varphi_1\cos\varphi_1\sin\varphi_2) \\
& + u_{x_2}u_{\varphi_1}(-2f_1\sin 2\varphi_1\cos\varphi_2 + 2f_2\cos^2\varphi_1\cos^2\varphi_2 \\
& + f_3\cos^2\varphi_1\sin 2\varphi_2)\sin\varphi_1 + u_{x_3}u_{\varphi_1}(-f_1\sin 2\varphi_1\sin\varphi_2 \\
& + f_2\cos^2\varphi_1\sin 2\varphi_2 + 2f_3\cos^2\varphi_1\sin^2\varphi_2)\sin\varphi_1 \\
& + u_{x_1}u_{\varphi_2}(f_2\sin\varphi_2 - f_3\cos\varphi_2)\sin\varphi_1 \\
& + u_{x_2}u_{\varphi_2}(f_1\sin\varphi_1\sin\varphi_2 - f_2\cos\varphi_1\sin 2\varphi_2 + f_3\cos\varphi_1\cos 2\varphi_2) \\
& + u_{x_3}u_{\varphi_2}(-f_1\sin\varphi_1\cos\varphi_2 + f_2\cos\varphi_1\cos 2\varphi_2 + f_3\cos\varphi_1\sin 2\varphi_2)\Big],
\end{aligned}$$

$$\mathrm{d}_5(u) = \frac{\partial}{\partial \varphi_2}\Big[-u_{x_1}u_{x_2}\cot\varphi_1\sin\varphi_2 - u_{x_2}^2\sin\varphi_2\cos\varphi_2$$

$$+ u_{x_2}u_{x_3}\cos 2\varphi_2 + u_{x_1}u_{x_3}\cot 2\varphi_1\cos\varphi_2$$

$$+ u_{x_3}^2\sin\varphi_2\cos\varphi_2 + u_{x_1}u_{\varphi_1}(f_2\sin\varphi_2 - f_3\cos\varphi_2)$$

$$+ u_{x_2}u_{\varphi_1}(u_1\sin\varphi_2 - f_2\cot\varphi_1\sin 2\varphi_2 + f_3\cot\varphi_1\cos 2\varphi_2)$$

$$+ u_{x_3}u_{\varphi_1}(-f_1\cos\varphi_2 + f_2\cot\varphi_1\cos 2\varphi_2 - f_3\cot\varphi_1\sin 2\varphi_2)$$

$$+ 2u_{x_2}u_{\varphi_2}(f_2\sin\varphi_2 - f_3\cos\varphi_2)\frac{\sin\varphi_2}{\sin^2\varphi_1}$$

$$+ 2u_{x_3}u_{\varphi_2}(f_3\cos\varphi_2 - f_2\sin\varphi_2)\frac{\sin\varphi_2}{\sin^2\varphi_1}\Big].$$

Proof. To prove Lemma 1.2.1, we transform each term in the expression for L_2L_1u so that the resulting expression contains only $u_{x_i}u_{x_j}$, $u_{\theta_k}u_{\theta_r}$, $u_{x_i}u_{\theta_k}$, and divergent terms, where $i,j = 1,2,3$ and $k,r = 1,2$.

Assume that the function $f = (f_1, f_2, f_3)$ is such that the quadratic form

$$\sum_{i,j=1}^{3} f_{ix_j}\xi^i\xi^j \geq \alpha|\xi|^2, \tag{1.2.3}$$

for any $\xi \in \mathbb{R}^3$, where α is a sufficiently large number depending only on f, f_{θ_i} $(i = 1,2)$, $|f_{ix_j} + f_{jx_i}|$ $(i \neq j,\ i,j = 1,2,3)$, and $d = |f_{2x_2} - f_{3x_3}|$. Moreover, suppose the following inequality holds:

$$J(u) \geq \alpha_1(|\nabla_x u|^2 + |\nabla_\theta u|^2), \quad \alpha_1 > 0. \tag{1.2.4}$$

Obviously, such functions exist. Indeed, it is easy to verify that the coefficient of $u_{\theta_1}^2$ in the expression for $J(u)$ is greater than α. To this end, it suffices to set $\xi_1 = -\sin\varphi_1$, $\xi_2 = \cos\varphi_1\cos\varphi_2$, and $\xi_3 = \cos\varphi_1\sin\varphi_2$ in inequality (1.2.3). One can prove analogously that the coefficient of $u_{\theta_2}^2$ in $J(u)$ is greater than α. In this case, it suffices to set $\xi_1 = 0$, $\xi_2 = -\sin\varphi_2$, and $\xi_3 = cos\varphi_2$. If f, f_{θ_i} $(i = 1,2)$, d, and $|f_{ix_j} + f_{jx_i}|$ $(i \neq j,\ i,j = 1,2,3)$ are sufficiently small compared to α, then (1.2.4) holds. □

Problem 1.2.1′. Find a pair of functions (u, λ) defined in the domain Ω satisfying the equation

$$L_1 u = \lambda + \overline{F}, \tag{1.2.1′}$$

provided that the functions f and $\overline{F}$ are given, the trace of the solution of equation (1.2.1′) on Γ is zero, and λ satisfies condition (1.2.2).

2. We now turn to the question of the uniqueness of the solution to Problem 1.2.1.

Theorem 1.2.1. *Let $f(x,\nu) = (f_1, f_2, f_3) \in C^2(\Omega)$ and assume that inequality (1.2.3) holds for any $\xi \in \mathbb{R}^3$, where α is chosen so that inequality (1.2.4) holds. Then Problem 1.2.1 has at most one solution u such that $u \in \Gamma(A)$ and $\lambda \in L_2(\Omega)$.*

Proof. Suppose that (u, λ) is a solution to the homogeneous version of Problem 1.2.1 such that $u \in \Gamma(A)$ and $\lambda \in L_2(\Omega)$. From (1.2.1) and (1.2.2) it follows that $Au = 0$. Hence there exists a sequence $\{u_k\} \subset \mathcal{M}^0$ such that

$$\begin{aligned} &1.\ u_k \to u \quad \text{weakly in } L_2(\Omega) \\ &2.\ \langle Au_k,\, u_k\rangle \to 0 \end{aligned} \tag{1.2.5}$$

as $k \to \infty$. The following equality is true:

$$-2\langle Au_k,\, u_k\rangle = \int_\Omega J(u_k)\, d\Omega. \tag{1.2.6}$$

Indeed, we have

$$\begin{aligned} -2Au_k u_k \equiv -2\Big[\Big(&-\sin\varphi_2\frac{\partial}{\partial x_2} + \cos\varphi\frac{\partial}{\partial x_3}\Big)\frac{\partial}{\partial\theta_2}(L_1u_k) \\ &+\Big(-\sin\varphi_1\frac{\partial}{\partial x_1} + \cos\varphi_1\cos\varphi_2\frac{\partial}{\partial x_2} \\ &+\cos\varphi_1\sin\varphi_2\frac{\partial}{\partial x_3}\Big)\frac{\partial}{\partial\theta_1}(L_1u_k)\Big]u_k \\ = 2\Big[u_k\frac{\partial}{\partial\theta_2}&(L_1u_k)\sin\varphi_2\Big]_{x_2} - 2\Big[u_k\frac{\partial}{\partial\theta_2}(L_1u_k)\cos\varphi_2\Big]_{x_3} \\ +2\Big[u_k\frac{\partial}{\partial\theta_1}&(L_1u_k)\sin\varphi_1\Big]_{x_1} - 2\Big[u_k\frac{\partial}{\partial\theta_1}(L_1u_k)\Big]_{x_2}\cos\varphi_1\cos\varphi_2 \\ -2\Big[u_k\frac{\partial}{\partial\theta_1}&(L_1u_k)\cos\varphi_1\sin\varphi_2\Big]_{x_3} + 2\frac{\partial}{\partial\theta_2}(L_1u_k)(\nabla_x u_k\cdot\nu'') \\ +2\frac{\partial}{\partial\theta_1}&(L_1u_k)(\nabla_x u_k\cdot\nu'). \end{aligned} \tag{1.2.7}$$

Then, from Lemma 1.2.1, and since $u_k \in \mathcal{M}^0$, integrating equality (1.2.7) over Ω we obtain

$$-2\langle Au_k, u_k\rangle = \int_\Omega \Big[J(u_k) + \sum_{i=1}^{5} \mathrm{d}_i(u_k)\Big]\, \mathrm{d}\Omega = \int_\Omega J(u_k)\, \mathrm{d}\Omega.$$

The divergent term $\sum_{i=1}^{5} \mathrm{d}_i(u_k)$ vanishes after integration over Ω since u_k is 2π-periodic in φ_2 and $u_k = 0$ on Γ and because of the presence of the multiplier $\sin\varphi_1$ in all terms that involve $\partial/\partial\varphi_1$ (except for several terms that include the derivatives of u_k with respect to φ_2). From the specific dependence of u_k on φ_2 it follows that these terms also contain the multiplier $\sin\varphi_1$ $(u_k(x, \cos\varphi_1, \sin\varphi_1\cos\varphi_2, \cos\varphi_1\sin\varphi_2))$.

Since the domain D is bounded and $u_k \in \mathcal{M}^0$, from (1.2.4) it follows that $\|u_k\|^2 \le C \int_\Omega J(u_k)\, \mathrm{d}\Omega$, where $C > 0$ does not depend on k. Then, as in the proof of Theorem 1.1.1, we have

$$\|u\|^2 \le \varliminf_{k\to\infty} \|u_k\|^2 \le -C \lim_{k\to\infty} \langle Au_k, u_k\rangle = 0,$$

i.e., $u = 0$, and (1.2.1) implies $\lambda = 0$. Theorem 1.2.1 is proved. □

3. We now turn to the proof of the existence theorem for Problem 1.2.1′.

Theorem 1.2.2. *Under the hypotheses of Theorem 1.2.1, suppose that* $\overline{F} \in \overline{H}_{1,2}(\Omega)$. *Then there exists a solution* (u, λ) *of Problem 1.2.1′ such that* $u \in \Gamma(A)$, $\lambda \in L_2(\Omega)$, *and*

$$\|u\|_{\overline{H}_1(\Omega)} + \|\lambda\| \le C\, \|\overline{F}\|_{\overline{H}_1(\Omega)},$$

where $C > 0$ *depends on* f *and the size of the domain* D.

Proof. Consider the following auxiliary problem:

$$Au \equiv \Big(-\sin\varphi_2\frac{\partial}{\partial x_2} + \cos\varphi_2\frac{\partial}{\partial x_3}\Big)\frac{\partial}{\partial\theta_2}(L_1u)$$

$$+ \Big(-\sin\varphi_1\frac{\partial}{\partial x_1} + \cos\varphi_1\sin\varphi_2\frac{\partial}{\partial x_3}\Big)\frac{\partial}{\partial\theta_1}(L_1u) = \mathcal{F} \qquad (1.2.8)$$

$$u\big|_\Gamma = 0, \qquad \Gamma = \partial D \times (0, \pi) \times (0, 2\pi), \qquad (1.2.9)$$

where $\mathcal{F} = \nu' \cdot \nabla_x \overline{F}_{\theta_1} + \nu'' \cdot \nabla_x \overline{F}_{\theta_2}$.

This problem will be referred to as the problem (1.2.8) – (1.2.9).

We shall seek its approximate solution u_N in the form

$$\sum_{i,n}^{N}\Big\{a_{0,i}^{(n)}P_n(\cos\varphi_1)+\sum_{m=1}^{n}\big[a_{m,i}^{(n)}\cos m\varphi_2+b_{m,i}^{(n)}\sin m\varphi_2\big]P_{n,m}(\cos\varphi_1)\Big\}\omega_i.$$

The unknown coefficients $a_{0,i}^{(n)}$, $a_{m,i}^{(n)}$, $b_{m,i}^{(n)}$ are determined from the following relations:

$$\langle Au_N-\mathcal{F},\ \omega_iP_n(\cos\varphi_1)\rangle=0, \tag{1.2.10}$$

$$\langle Au_N-\mathcal{F},\ \omega_iP_{n,m}(\cos\varphi_1)\cos m\varphi_2\rangle=0, \tag{1.2.11}$$

$$\langle Au_N-\mathcal{F},\ \omega_iP_{n,m}(\cos\varphi_1)\sin m\varphi_2\rangle=0, \tag{1.2.12}$$

where $i,n=1,2,\dots,N$; $m=1,2,\dots,n$.

It is easy to see that the total number of relations in (1.2.10) – (1.2.12) is equal to the number of the unknown coefficients $a_{0,i}^{(n)}$, $a_{m,i}^{(n)}$, $b_{m,i}^{(n)}$, namely, $l(N)=N^2+\sum\limits_{n=1}^{n}n(n+1)$.

We now prove that under the assumptions of Theorem 1.2.2 the system of linear algebraic equations (1.2.10) – (1.2.12) has a unique solution $u_N=(a_{0,i}^{(n)},\ a_{m,i}^{(n)},\ b_{m,i}^{(n)})$, $i,n=1,2,\dots,N$, $m=1,2,\dots,n$, for every function $\mathcal{F}\in L_2(\Omega)$. For this purpose, we consider the homogeneous version of the system (1.2.10) – (1.2.12) (with $\mathcal{F}=0$). Multiply equation (i,n) of (1.2.10) by $-2a_{0,i}^{(n)}$, equation (i,n,m) of (1.2.11) by $-2a_{m,i}^{(n)}$, and equation (i,n,m) of (1.2.12) by $-2b_{m,i}^{(n)}$. Summing the equations from 1 to N with respect to (i,n), and from 1 to n with respect to m, we obtain

$$-2\langle Au_N,\ u_N\rangle=0.$$

Then equality (1.2.6) yields

$$\int_\Omega J(u_N)\,\mathrm{d}\Omega=0. \tag{1.2.13}$$

By the assumption of Theorem 1.2.2, $J(u_N)\geq\alpha_1(|\nabla_x u_N|^2+|\nabla_\theta u_N|^2)$. Hence, (1.2.13) and the condition $u_N=0$ on Γ imply that $u_N=0$ on Ω. Since the system $\{\omega_iP_n,\ \omega_i\cos m\varphi_2P_{n,m}(\cos\varphi_1),\ \omega_i\sin m\varphi_2P_{n,m}(\cos\varphi_1)\}$ $(i,n=1,2,3,\dots;\ m=1,2,3,\dots,n)$ is linearly independent, we have $a_{0,i}^{(n)}=a_{m,i}^{(n)}=b_{m,i}^{(n)}=0$ $(i,n=1,2,\dots,N;\ m=1,2,\dots,n)$.

Thus, the homogeneous version of the system of linear algebraic equations (1.2.10) – (1.2.12) has only trivial solution and therefore the original inhomogeneous system (1.2.10) – (1.2.12) has a unique solution u_N for every $\mathcal{F} \in L_2(\Omega)$.

Next, estimate u_N in terms of the right-hand side $\mathcal{F}$ of equation (1.2.8), where u_N is determined by the coefficients $a_{0,i}^{(n)}$, $a_{m,i}^{(n)}$, $b_{m,i}^{(n)}$ after solving the system (1.2.10) – (1.2.12). To this end, multiply equation (i, n) of (1.2.10) by $-2a_{0,i}^{(n)}$, equation (i, n, m) of (1.2.11) by $-2a_{m,i}^{(n)}$, and equation (i, n, m) of (1.2.12) by $-2b_{m,i}^{(n)}$. Summing the equations from 1 to N with respect to (i, n), and from 1 to n with respect to m, we obtain

$$-2\langle Au_N,\, u_N\rangle = -2\langle F,\, u_N\rangle. \tag{1.2.14}$$

Observing that $u_N = 0$ on Γ, we can estimate the right-hand side of the obtained equality as follows:

$$-2\int_\Omega \mathcal{F} u_N \,\mathrm{d}\Omega \le \beta \int_\Omega |\nabla_x \overline{F}|^2 \,\mathrm{d}\Omega + \beta^{-1} \int_\Omega |\nabla_\theta u_N|^2 \,\mathrm{d}\Omega,$$

where $\beta^{-1} < \alpha$ and $\mathcal{F} = \nu' \cdot \nabla_x \overline{F}_{\theta_1} + \nu'' \cdot \nabla_x \overline{F}_{\theta_2}$.

As was shown above, the left-hand side of (1.2.14) equals $\int_\Omega J(u_N)\,\mathrm{d}\Omega$. Hence (1.2.14) implies

$$\int_\Omega (|\nabla_x u_N|^2 + |\nabla_\theta u_N|^2)\,\mathrm{d}\Omega \le \int_\Omega J(u_N)\,\mathrm{d}\Omega$$

$$\le \beta \int_\Omega |\nabla_x \overline{F}|^2 \,\mathrm{d}\Omega + \beta^{-1} \int_\Omega |\nabla_\theta u|^2 \,\mathrm{d}\Omega.$$

Since Ω is bounded and $u_N = 0$ on Γ, the foregoing inequality yields

$$\|u_N\|_{\overline{H}_1^0(\Omega)} \le C \,\|\, |\nabla_x \overline{F}| \,\|,$$

where the constant C does not depend on N.

Using the last inequality, we can prove that there exists a sequence $\{u_N\}$ such that u_{Nx_i} and $u_{N\theta_j}$ ($i = 1, 2, 3,\ j = 1, 2$) weakly converge in $L_2(\Omega)$ to u_{x_i} and u_{θ_j}, respectively, as $N \to \infty$, where $u \in \overline{H}_1^0(\Omega)$ is the weak limit of the sequence $\{u_N\}$ in $\overline{H}_1^0(\Omega)$. Moreover, $\|u\|_{\overline{H}_1(\Omega)} \le C\, \|\overline{F}\|_{\overline{H}_1(\Omega)}$.

These facts are established in exactly the same way as in the proof of Theorem 1.1.2 using estimate (1.1.16) from Section 1.1.

In equalities (1.2.10) – (1.2.12), we transfer the appropriate differentiation operators to the functions

$$P_{n,m}(\cos\varphi_1)\cos m\varphi_2, \quad P_{n,m}(\cos\varphi_1)\sin m\varphi_2, \quad P_n(\cos\varphi_1).$$

Since $u_N = 0$ on Γ and $\mathrm{d}\Omega = \sin\varphi_1\,\mathrm{d}D\,\mathrm{d}\varphi_1\,\mathrm{d}\varphi_2$, for $N \geq \bar{n} \geq \bar{m}$ we have

$$\Big\langle L_1 u - \overline{F},\ \Big[\frac{\partial}{\partial\theta_1}(\nabla_x r^{(l)}_{\bar{n},\bar{m}} \cdot \nu') + \frac{\partial}{\partial\theta_2}(\nabla_x r^{(l)}_{\bar{n},\bar{m}} \cdot \nu'') + (\nabla_x r^{(l)}_{\bar{n},\bar{m}} \cdot \nu')\cot\varphi_1\Big]\Big\rangle = 0, \tag{1.2.15}$$

where $l = 1,2$; $r^{(1)}_{\bar{n},\bar{m}} = P_{\bar{n},\bar{m}}(\cos\varphi_1)\sin\bar{m}\varphi_2$, $r^{(2)}_{\bar{n},\bar{m}} = P_{\bar{n},\bar{m}}(\cos\varphi_1)\cos\bar{m}\varphi_2$, $\bar{m} = 0,1,2,\dots,\bar{n}$, $P_{\bar{n},0} = P_{\bar{n}}$.

Since the sequences of u_N and its first-order derivatives with respect to x_i and θ_j $(i = 1,2,3,\ j = 1,2)$ weakly converge in $L_2(\Omega)$ as $N \to \infty$, we can pass to the limit as $N \to \infty$ in equality (1.2.15).

Because the linear span of the set of functions $r^{(l)}_{\bar{n},\bar{m}}$ is everywhere dense in $\overline{H}^0_{1,2}(\Omega)$, passing to the limit as $N \to \infty$ in (1.2.15), we get

$$\Big\langle L_1 u - \overline{F},\ \Big[\frac{\partial}{\partial\theta_1}(\nabla_x \eta \cdot \nu') + \frac{\partial}{\partial\theta_2}(\nabla_x \eta \cdot \nu'') + (\nabla_x \eta \cdot \nu')\cot\varphi_1\Big]\Big\rangle = 0$$

for any $\eta \in \overline{H}^0_{1,2}(\Omega)$.

The final part of the proof of Theorem 1.2.2 is the same as that of Theorem 1.1.2. □

Remark 1.2.2. The results analogous to Theorems 1.1.1 – 1.2.2 can be proved for Problems 1.1.1 and 1.2.1 if we set $f = 0$ and $F = (F_1, F_2) = 0$, respectively. Note that the results formulated in Section 1.2 are valid also for the transport equation of the general form:

$$L_1 u + \beta u + \int_0^{\pi}\int_0^{2\pi} K(x,\varphi,\varphi')u(x,\varphi')\,\mathrm{d}\varphi' = \lambda(x,\varphi),$$

where the given functions β and K are sufficiently smooth. In this case, α_1 will depend also on these functions.

1.3. SOLVABILITY OF THE PROBLEM OF INTEGRAL GEOMETRY ALONG GEODESICS

1. In this section we study the solvability of the inverse problem for a special kinetic equation and the related integral geometry problem along geodesics. We prove the theorems of existence and uniqueness of solutions to the problems in question under the assumption that a certain quadratic form is of constant sign (more precisely, that the sectional curvature of the Riemannian space associated with the kinetic equation is nonpositive).

We introduce the following notation. Let D be a domain in the real Euclidean space $\mathbb{R}^n$ $(n \geq 1)$ of variables x, with boundary $\partial D \in C^1$, and

$$\Omega = \{(x,\xi) \,:\, x \in D,\ \xi = (\xi^1, \dots, \xi^n),\ g_{ij}\xi^i\xi^j < 1\},$$

where g_{ij} is a given metric on D, which determines a Riemannian manifold $\mathcal{M}$, such that $g_{ij} \in C^2(\overline{D})$, $g_{ij} = g_{ji}$, $i, j = 1, 2, \dots, n$. Here and in what follows, whenever an index appears in an expression both as a superscript and a subscript, summation from 1 to n is assumed with respect to this index. The boundary of Ω is represented in the form $\partial\Omega = \Gamma_1 \cup \Gamma_2$, where

$$\Gamma_1 = \{(x,\xi) \,:\, x \in \partial D,\ g_{ij}\xi_i\xi_j < 1\}, \quad \Gamma_2 = \partial\Omega \setminus \Gamma_1,$$

We also set

$$\Omega_\delta = \{(x,\xi) \,:\, x \in D,\ \delta < g_{ij}\xi^i\xi_j < 1\},$$

$$\Gamma_{1\delta} = \{(x,\xi) \,:\, x \in \Gamma_1,\ \delta < g_{ij}\xi^i\xi_j \leq 1\},$$

$$\Gamma_{2\delta} = \{(x,\xi) \,:\, x \in D,\ g_{ij}\xi^i\xi_j = \delta\}.$$

By $H_{1,2}(\Omega)$ we denote the set of all real-valued functions $u(x,\xi) \in L^2(\Omega)$ that have generalized derivatives u_{x^i}, u_{ξ^i}, $u_{x^i\xi^j}$, $u_{\xi^i\xi^j}$ $(i, j = 1, 2, \dots, n)$, which belong to $L_2(\Omega)$.

2. Consider the equation

$$Lu \equiv \xi^i \frac{\partial u}{\partial x^i} - \Gamma^i_{jk}\xi^i\xi^k \frac{\partial u}{\partial \xi^i} = \lambda(x,\xi) \tag{1.3.1}$$

in the domain Ω with the right-hand side λ such that the following equality holds for any function $\eta \in H_{1,2}(\Omega)$ whose trace on Γ_1 is zero:

$$\int_\Omega g^{i,j}\lambda\nabla'_j \frac{\partial \eta}{\partial \xi^i}\,d\Omega - \int_{\partial\Omega} g g^{i,j}\lambda\nabla'_j\eta\,\bar{\mathbf{n}}_{\xi^i}\,dS = 0, \tag{1.3.2}$$

$$\text{where} \quad \nabla'_j = \nabla_j - \Gamma^i_{jk}\xi^k \frac{\partial}{\partial \xi^i}$$

and ∇_j is the operator of covariant differentiation; the matrix (g^{ij}) is the inverse of (g_{ij}), g is the determinant of (g_{ij}), $\mathrm{d}\Omega = g\,\mathrm{d}x\,\mathrm{d}\xi$, Γ^i_{jk} is the symmetric connection compatible with the metric g_{ij} (Christoffel symbols) (see Dubrovin, Novikov, and Fomenko, 1979), and $\bar{\mathbf{n}}_{\xi^i}$ is the cosine of the angle made by the outward normal $\bar{\mathbf{n}}$ to $\partial\Omega$ and the axis ξ^i.

Remark 1.3.1. Equality (1.3.2) implies that the function λ satisfies the equation

$$g^{ij}\nabla'_j \frac{\partial}{\partial \xi^i}\lambda = 0$$

on Ω in the generalized sense. If $\lambda(x,\xi)$ does not depend on ξ, then λ satisfies condition (1.3.2).

Assume that the condition

$$-R^{is}_{rl}\,\xi^r\xi^l\eta_i\eta_s \geq 0 \tag{1.3.3}$$

holds for all $x \in \overline{D}^*$ and $\xi, \eta \in \mathbb{R}^n$, where $R^{is}_{rl} = g^{js}R^i_{jrl}$ and R^i_{rlj} are the components of the curvature tensor of the metric g_{ij} (the Riemannian curvature), $\overline{D}^*$ is the closed convex hull of the domain D.

Problem 1.3.1. Find a pair of functions (u,λ) defined in Ω from equation (1.3.1), provided that the trace of the solution $u(x,\xi)$ of equation (1.3.1) on Γ_1 is known:

$$u\big|_{\Gamma_1} = u_0(x,\xi).$$

Problem 1.3.2. Find a pair of functions (u,λ) defined in Ω_0 from equation (1.3.1), provided that the trace of the solution $u(x,\xi)$ of equation (1.3.1) on Γ_{10} is known.

Problem 1.3.3. Let a function $\lambda(x,\xi)$ satisfy (1.3.2) on Ω_0. Given the integrals of $\lambda(x,\xi)$ along the geodesics of $\mathcal{M}$ with endpoint $x \in \partial D$ and direction ξ at x such that $0 < g_{ij}\xi^i\xi^j < 1$, determine $\lambda(x,\xi)$ in Ω_0.

Let us explain how Problems 1.3.2 and 1.3.3 are interrelated.

Consider the Cauchy problem for the system

$$\ddot{z}^i = -\Gamma^i_{jk}\dot{z}^j\dot{z}^k \tag{1.3.4}$$

with the data

$$z^i(0) = z_0^i, \quad \dot{z}^i(0) = \dot{z}_0^i = \xi_0^i, \quad i = 1, 2, \dots, n, \tag{1.3.5}$$

where $\dot{z}^i = \mathrm{d}z^i/\mathrm{d}t$.

System (1.3.4) coincides with the equation for the geodesics of $\mathcal{M}$ under the assumption that the parameter of a curve is chosen to be proportional to the natural one (see Dubrovin, Novikov, and Fomenko, 1979). It is possible to choose such parameter if the tangent vector of the curve is nonzero. Thus, we shall assume that $|\dot{z}| \neq 0$ in (1.3.4).

If condition (1.3.3) holds, then the space $\mathcal{M}$ is scattering, i.e., every geodesic eventually leaves any compact set (see Gromoll, Klingenberg, and Meyer, 1975).

We introduce a function $u(x, \xi)$ as follows. Take a geodesic $\gamma(x, \xi)$ with endpoint $x \in D$ and "direction" ξ at x. The geodesic $\gamma(x, \xi)$ is the projection of the solution of the Cauchy problem (1.3.4) – (1.3.5) with the data $z_0^i = x^i$, $\xi_0^i = \xi^i$, $x = (x^1, x^2, \dots, x^n)$, $\xi = (\xi^1, \xi^2, \dots, \xi^n)$, onto the domain D of the space X. By assumption, this geodesic intersects the boundary of D. Write the integral of λ over the part of $\gamma(x, \xi)$ that lies in D (denoted by $\gamma_D(x, \xi)$):

$$u = \int_{\gamma_D(x,\xi)} \lambda \, \mathrm{d}S, \tag{1.3.6}$$

where λ is the same function as in Problem 1.3.2 and $\mathrm{d}S$ is the arc length element of the geodesic γ_D. Differentiating (1.3.6) at the point x in the "direction" ξ, i.e., differentiating (1.3.6) with respect to the parameter t and taking into account that $\gamma(x, \xi)$ is the solution of the Cauchy problem (1.3.4) – (1.3.5) in D with the data x^i, ξ^i, we easily obtain the following result: the pair (u, λ) satisfies equation (1.3.1) in the domain Ω_0 and from the problem statement it follows that $u(x, \xi)$ is known on Γ_{10}, i.e., we arrive at Problem 1.3.2.

Problem 1.3.3 is a problem of integral geometry for the geodesics of the space $\mathcal{M}$. The information about the solution to Problem 1.3.3 is given only on Γ_{10} and therefore we can determine the solution only on Ω_0. Consequently, we have to show in the proof of the existence theorem that the function $\lambda(x, \xi)$ (and hence $Lu(x, \xi)$) has a trace on $\partial\Omega_0$.

Let $\widetilde{C}_0^3(\Omega, \Gamma_1)$ be the set of all real-valued functions $u(x, \xi) \in C^3(\Omega)$ such that $u = 0$ on Γ_1. Introduce a scalar product in $\widetilde{C}_0^3(\Omega, \Gamma_1)$:

$$(u, z)_{\nabla'} = \langle u, z \rangle + \sum_{i=1}^{n} \langle \nabla_i' u, \nabla_i' z \rangle,$$

where $\langle .,. \rangle$ is the scalar product in $L_2(\Omega)$. Let $H(\nabla')$ denote the closure of the set $\widetilde{C}_0^3(\Omega, \Gamma_1)$ with respect to the norm $\|u\|^2_{\nabla'} = (u, u)_{\nabla'}$, and put

$$Au \equiv g^{ij} \nabla'_j \frac{\partial}{\partial \xi^i}(Lu).$$

By $\Gamma''(A)$ we denote the set of functions $u \in H(\nabla')$ with the property that for any $u \in \Gamma''(A)$ there exists a function $y \in L_2(\Omega)$ such that $\forall \eta \in \widetilde{C}_0^3(\Omega, \Gamma_1)$

$$\int_\Omega Lu\, g^{ij} \nabla'_j \frac{\partial \eta}{\partial \xi^i}\, d\Omega - \int_{\Gamma_2} g g^{ij}\, Lu\, \nabla'_j \eta\, \bar{\mathbf{n}}_{\xi^i}\, dS = \int_\Omega y\, \eta\, d\Omega$$

and $y = Au$.

Assume that $\Gamma(A) \subset \Gamma''(A)$ is such that for any $u \in \Gamma(A)$ there is a sequence $\{u_k\} \subset \widetilde{C}_0^3(\Omega, \Gamma_1)$ with the following properties:

1. $u_k \to u$ weakly in $H(\nabla')$
2. $\langle Au_k, u_k \rangle \to \langle Au, u \rangle$

as $k \to \infty$. Let $\overset{\circ}{H}_{\overline{1},2}(\Omega) = H_{1,2}(\Omega) \cap \overset{\circ}{H}_1(\Omega, \Gamma_1)$ and let $\Gamma'(A)$ be the closure of $\widetilde{C}_0^3(\Omega, \Gamma_1)$ with respect to the norm $\|u\|_{\Gamma(A)} = \|u\| + \|Au\|$, where $\|\cdot\|$ is the norm in $L_2(\Omega)$ with weight $g(x)$.

The following inclusions hold:

$$\Gamma'(A) \cap H(\nabla') \subset \Gamma(A) \subset \Gamma''(A),$$

$$\Gamma''(A) \cap (H_2(\Omega) \cap \overset{\circ}{H}_1(\Omega, \Gamma_1)) \subset \Gamma(A) \subset L_2(\Omega).$$

Let $\{\omega_1, \omega_2, \omega_3, \dots\}$ be a complete orthonormalized set in $L_2(\Omega)$ with weight $g(x)$, where $\omega_i \in \widetilde{C}_0^3(\Omega, \Gamma_1)$, $i = 1, 2, \dots$. Assume that the linear span of this set is everywhere dense in $\overset{\circ}{H}_{\overline{1},2}(\Omega)$ and $\mathcal{P}_n$ is the orthogonal projector of $L^2(\Omega)$ onto $\mathfrak{M}_n$, where $\mathfrak{M}_n$ denotes the linear span of the set $\{\omega_1, \omega_2, \omega_3, \dots, \omega_n\}$.

Consider the following problem.

Problem 1.3.1′. Find a pair of functions (u, λ) defined in Ω and satisfying the equation

$$Lu = \lambda + F, \qquad (1.3.1')$$

provided that F is given, the trace of the solution of (1.3.1′) on Γ' is zero, and λ satisfies condition (1.3.2).

As in Section 1.1, one can prove that if $\partial D \in C^{[n/2]+4}$ and $u_0 \in H^{5/2}(\Gamma_1)$, then Problem 1.3.1 can be reduced to Problem 1.3.1′ with $F \in H^2(\Omega)$.

3. In this part of the section we prove Theorem 1.3.1.

Theorem 1.3.1. *Provided that condition (1.3.3) holds, Problem* 1.3.1′ *has at most one solution* (u, λ) *such that* $u \in \Gamma(A)$ *and* $\lambda \in L_2(\Omega)$.

Before proving Theorem 1.3.1, we present some auxiliary facts that will be necessary in the sequel. In this section, by the symbol ∇_j in the definition of the operator ∇'_j ($\nabla'_j = \nabla_j - \xi^k \Gamma^i_{jk}\, \partial/\partial\xi^i$) we mean the operator of covariant differentiation. As is known, it is defined as follows (see Dubrovin, Novikov, and Fomenko, 1979):

$$\begin{aligned} &\nabla_j u = \frac{\partial u}{\partial x_j}, && \text{for a function} \\ &\nabla_j u^i = \frac{\partial u^i}{\partial x_j} + \Gamma^i_{jk} u^k, && \text{for a vector} \\ &\nabla_j u_i = \frac{\partial u_i}{\partial x_j} - \Gamma^k_{ji} u_k, && \text{for a covector} \\ &\nabla_j u^i_k = \frac{\partial u^i_k}{\partial x_j} + \Gamma^i_{sj} u^s_k - \Gamma^s_{ki} u^i_s, && \text{for a tensor} \end{aligned} \tag{1.3.7}$$

Using (1.3.7), one can easily obtain

$$\begin{aligned} &\nabla'_j \xi^i = 0, \qquad \frac{\partial}{\partial \xi^i} \nabla'_j u = \nabla'_j \frac{\partial u}{\partial \xi^i}, \\ &(\nabla'_i \nabla'_j - \nabla'_j \nabla'_i) u = R^s_{ikj} \xi^k \frac{\partial u}{\partial \xi^s}. \end{aligned} \tag{1.3.8}$$

Furthermore, the following equalities hold:

$$\nabla'_k g_{ij} = 0, \quad \nabla'_k g^{ij} = 0. \tag{1.3.9}$$

The first equality in (1.3.9) holds because the connection Γ^k_{ij} is compatible with the metric g_{ij} (see Dubrovin, Novikov, and Fomenko, 1979). The second equality can be proved using the relation $\nabla'_k \delta^i_j = 0$, where δ^i_j is the Kronecker delta.

Proof. Suppose that (u, λ) is a solution to the homogeneous version of Problem 1.3.1 such that $u \in \Gamma(A)$ and $\lambda \in L_2(\Omega)$. From (1.3.1) and (1.3.2) it follows that $Au = 0$; therefore, there exists a sequence $\{u_k\} \subset \widetilde{C}_0^3(\Omega, \Gamma_1)$ such that, as $k \to \infty$,

1. $u_k \to u$ weakly in $H(\nabla')$
2. $\langle Au_k, u_k\rangle \to 0$

Furthermore, the following equality is valid

$$-2\langle Au_k, u_k\rangle = J(u_k), \tag{1.3.10}$$

where

$$J(u_k) = \int_\Omega \Big(g^{ij}\,\nabla'_i u_k\,\nabla'_j u_k - R^{si}_{rl}\,\xi^r\xi^l\,\frac{\partial u_k}{\partial\xi^s}\frac{\partial u_k}{\partial\xi^i}\Big)\,d\Omega + \int_{\partial\Omega}(Lu_k)^2\frac{g\,dS}{\Big[\sum\limits_{j=1}^{n}(g_{ij}\xi^i)^2\Big]^{1/2}}.$$

Indeed, we have

$$-2g^{ij}\,\nabla'_j\frac{\partial}{\partial\xi^i}(Lu_k)u_k = 2\nabla'_j\Big[g^{ij}\frac{\partial}{\partial\xi^i}(Lu_k u_k)\Big] + 2g^{ij}\,\frac{\partial}{\partial\xi^i}(Lu_k)\nabla'_j u_k. \tag{1.3.11}$$

We now prove the equality

$$2\int_\Omega g^{ij}\,\frac{\partial}{\partial\xi^i}(Lu_k)\nabla'_j u_k\,d\Omega = \int_{\partial\Omega} g^{ij}\nabla'_i u_k\,Lu_k\,g\,\mathbf{n}_{\xi^i}\,dS + \int_\Omega g^{ij}\,\nabla'_j u_k\nabla'_i u_k\,d\Omega - \int_\Omega R^{si}_{rl}\,\xi^r\xi^l\frac{\partial u_k}{\partial\xi^i}\frac{\partial u_k}{\partial\xi^s}\,d\Omega \tag{1.3.12}$$

The following identity holds:

$$\begin{aligned}2g^{ij}(Lu_k)_{\xi^i}\,\nabla'_j u_k &= g^{ij}\nabla'_i u_k\nabla'_j u_k - R^{s}_{rlj}\,\xi^r\xi^l g^{ij}\frac{\partial u_k}{\partial\xi^s}\frac{\partial u_k}{\partial\xi^i} + (g^{ij}\nabla'_i u_k\,Lu_k)_{\xi^j}\\ &\quad + \nabla'_l\Big(g^{ij}\nabla'_j u_k + \frac{\partial u_k}{\partial\xi^i}\xi^l\Big) - \nabla'_j\Big(g^{ij}\xi^l\frac{\partial u_k}{\partial\xi^i}\nabla'_l u_k\Big).\end{aligned} \tag{1.3.13}$$

Indeed, by (1.3.7) – (1.3.9), we have

$$2g^{ij}\nabla'_j u_k \frac{\partial}{\partial \xi^i}(Lu_k)$$

$$= 2g^{ij}\nabla'_j u_k \nabla'_i u_k + g^{ij}\nabla'_j u_k\Big(\xi^l \frac{\partial}{\partial \xi^i}\nabla'_l u_k + \xi^l \nabla'_l \frac{\partial u_k}{\partial \xi^i}\Big)$$

$$= g^{ij}\nabla'_j u_k \nabla'_i u_k + \frac{\partial}{\partial \xi^i}(g^{ij}\nabla'_j u_k \nabla'_l u_k \xi^l) - g^{ij}\xi^l \nabla'_j \frac{\partial u_k}{\partial \xi^i}\nabla'_l u_k$$

$$+ \nabla'_l\Big(g^{ij}\xi^l \frac{\partial u_k}{\partial \xi^i}\nabla'_j u_k\Big) - g^{ij}\xi^l \frac{\partial u_k}{\partial \xi^i}\nabla'_l \nabla'_j u_k. \qquad (1.3.14)$$

On the other hand,

$$-g^{ij}\xi^l \frac{\partial u_k}{\partial \xi^i}\nabla'_l \nabla'_j u_k$$

$$= -g^{ij}\xi^l \frac{\partial u_k}{\partial \xi^i}\nabla'_j \nabla'_l u_k + g^{ij}\xi^l \frac{\partial u_k}{\partial \xi^i}(\nabla'_j \nabla'_l - \nabla'_j \nabla'_l)u_k \qquad (1.3.15)$$

$$= -\nabla'_j\Big(g^{ij}\xi^l \frac{\partial u_k}{\partial \xi^i}\nabla'_l u_k\Big) + g^{ij}\xi^l \nabla'_j \frac{\partial u_k}{\partial \xi^i}\nabla'_l u_k - g^{ij}\xi^l \frac{\partial u_k}{\partial \xi^i} R^s_{rlj}\,\xi^r \frac{\partial u_k}{\partial \xi^s}$$

Equalities (1.3.14) and (1.3.15) imply (1.3.13).

We introduce the following notation:

$$p(x,\xi) = 1 - g_{ij}\xi^i\xi^j, \quad \Omega_x = \{\xi \in \mathbb{R}^n : p(x,\xi) > 0\}, \quad d\xi = d\xi_1\, d\xi_2, \dots, d\xi_2.$$

For any vector function $w = (w^j(x,\xi)) \in C^1(\overline{\Omega})$, the following equality is true:

$$\sqrt{g}\int_{\Omega_x} = \nabla_j\Big[\sqrt{g}\int_{\Omega_x} w^j(x,\xi)\, d\xi\Big]. \qquad (1.3.16)$$

Indeed, we have

$$\nabla_j\Big[\sqrt{g}\int_{\Omega_x} w^j(x,\xi)\, d\xi\Big] = \nabla_j\Big[\sqrt{g}\int_{\mathbb{R}^n} w^j(x,\xi)\,\theta(p(x,\xi))\, d\xi\Big]$$

$$= \nabla_j(\sqrt{g})\int_{\mathbb{R}^n} w^j\,\theta(p)\, d\xi + \sqrt{g}\int_{\mathbb{R}^n}\nabla_j(w^j)\,\theta(p)\, d\xi$$

$$+ \sqrt{g}\int_{\partial\Omega_x} w^j p_{x^j}\, dS, \qquad (1.3.17)$$

where $\mathrm{d}S$ is the surface element of $\partial\Omega_x$, $p_{x^j} = -(g_{is})_{x^j}\xi^i\xi^s$, and $\theta(p)$ is the Heaviside function.

Moreover,

$$\begin{aligned}
\sqrt{g}\int_{\mathbb{R}^n} \nabla_j(w^j)\,\theta(p)\,\mathrm{d}\xi &= \sqrt{g}\int_{\mathbb{R}^n}\Big(\nabla'_j + \Gamma^i_{jk}\,\xi^k\frac{\partial}{\partial\xi^i}\Big)w^j\,\theta(p)\,\mathrm{d}\xi \\
&= \sqrt{g}\int_{\mathbb{R}^n}\nabla'_j(w^j\,\theta(p))\,\mathrm{d}\xi + \sqrt{g}\int_{\mathbb{R}^n}\frac{\partial}{\partial\xi^i}(\Gamma^i_{jk}\,\xi^k w^j\,\theta(p))\,\mathrm{d}\xi \\
&\quad - \sqrt{g}\,\Gamma^i_{ji}\int_{\mathbb{R}^n} w^j\,\theta(p)\,\mathrm{d}\xi + \sqrt{g}\int_{\partial\Omega_x}\Gamma^i_{jk}w^j g_{is}\,\xi^s\xi^k\,\mathrm{d}S.
\end{aligned}$$

Substituting the foregoing expression for $\sqrt{g}\int_{\partial\Omega_x}\nabla_j w^j\,\mathrm{d}\xi$ into (1.3.17) and using the relations

$$\Gamma^i_{ji} = \frac{1}{2g}\frac{\partial}{\partial x^j}(g),\quad \Gamma^i_{jk}g_{is} = \Gamma_{s,jk},\quad \Gamma_{i,kj} + \Gamma_{k,ij} = \frac{\partial g_{ik}}{\partial x^j}$$

(see Dubrovin, Novikov, and Fomenko, 1979), we obtain (1.3.16).

We multiply both sides of (1.3.13) by $g(x)$ and integrate over Ω taking into account (1.3.16), the condition $u = 0$ on Γ_1, and the relation

$$\nabla_i T^i = \frac{1}{\sqrt{g}}\frac{\partial}{\partial x^i}(\sqrt{g}T^i).$$

As a result, we have (1.3.12), from which we obtain (1.3.11) in exactly the same way, observing that

$$\eta_{\xi^i} = \frac{g_{ij}\xi^j}{\sqrt{\sum\limits_{j=1}^{n}(g_{ij}\xi^i)^2}}.$$

Since $u_k \to u$ weakly in $H(\nabla')$ and $\langle Au_k, u_k\rangle \to 0$ as $k\to\infty$, and

$$\int_\Omega g^{ij}\nabla'_j u_k\nabla'_i u_k\,\mathrm{d}\Omega \le J(u_k),$$

we have

$$\int_\Omega g^{ij}\nabla'_j u\,\nabla'_i u\,\mathrm{d}\Omega \le \varliminf_{k\to\infty}\int_\Omega g^{ij}\nabla'_j u_k\nabla'_i u_k\,\mathrm{d}\Omega \le \varliminf_{k\to\infty} J(u_k) = 0$$

and therefore $\nabla'_j u = 0$ for $j = 1, 2, \dots, n$, which yields $\lambda = 0$ in Ω because $\xi^j \nabla'_j u = \lambda = Lu$. Since $L = \mathrm{d}/\mathrm{d}s$ on a geodesic and $u = 0$ on Γ_1, it is easy to prove that $u = 0$ in Ω. □

We have proved the theorem for Problem 1.3.1. The proof of the same assertion for Problem 1.3.2 is similar. Note that in Problem 1.3.2 the trace of the solution is given on $\Gamma_{1\delta}$ and therefore $u = \lambda = 0$ is readily established only on Ω_δ. We note also that in equality (1.3.12) with Ω replaced by Ω_δ the integral over the part of $\partial\Omega_\delta$ of the form $\Gamma_{2\delta}$ tends to zero as $\delta \to 0$.

Remark 1.3.2. The uniqueness of solutions to the problems considered above essentially depends on the geometry of the corresponding domain. The domain Ω in Problem 1.3.1 is a "ball" bundle in $\mathcal{M}$. However, the assertion of Theorem 1.3.1 will not hold if for Ω we take a ball of the form $\widetilde{\Omega} = \{(x, \xi) : x \in \mathbb{R}^1, \xi \in \mathbb{R}^1, x^2 + (2 - \xi)^2 < 1\}$ in the Euclidean space $\mathbb{R}^2$. To prove this, consider the equation $\xi u_x = \lambda(x, \xi)$ in $\widetilde{\Omega}$. This equation satisfies condition (1.3.3). It can be directly verified that the pair of functions $u = \dfrac{1}{2\xi}[x^2 + (2 - \xi)^2 - 1]$ and $\lambda = x$ satisfies the equation $\xi u_x = \lambda$, $u = 0$ on $\partial\Omega$, and λ satisfies (1.3.2).

4. In the final part of the section, we formulate the following theorem.

Theorem 1.3.2. *Assume that a family of geodesics of the space $\mathcal{M}$ is regular in a domain D, $\partial D \in C^{[n/2]+4}$, $F \in H^2(\Omega)$, and inequality (1.3.3) holds. Then there exist functions $u \in \Gamma(A)$ and $\lambda \in L_2(\Omega)$ with the following properties:*

1) *$Lu \in L_2(\Omega)$, the pair (u, λ) satisfies equation* (1.3.1) *on Ω, and $u(x, \xi)$ satisfies the condition $u = 0$ on Γ_1;*

2) *the functions Lu and λ have "traces" on Γ_2;*

3) *λ satisfies condition* (1.3.2) *on Ω.*

Remark 1.3.3. In general, the trace on Γ_2 is not defined for functions from L_2. However, for $Lu, \lambda \in L_2(\Omega)$ in Theorem 1.3.2 there are functions in $L_2(\Gamma_2)$ that coincide with the traces of Lu and λ on Γ_2 if these traces exist. These are the functions that are called the "traces" of Lu and λ on Γ_2 in item 2) of Theorem 1.3.2.

Proof. Consider the following problem.

Find a solution of the equation

$$Au = \mathcal{F}, \tag{1.3.18}$$

provided that

$$u\big|_{\Gamma_1} = 0, \tag{1.3.19}$$

where

$$\mathcal{F} = g^{ij}\nabla'_j \frac{\partial}{\partial \xi^i} F.$$

We shall seek an approximate solution to (1.3.18) – (1.3.19) in the form

$$u_N = \sum_{i=1}^{N} \alpha_{Ni}\omega_i(x, \xi)$$

using the relations

$$\int_\Omega g^{ij}\nabla'_j \frac{\partial}{\partial \xi^i}(Lu_N - F)\omega_l \, d\Omega = 0, \tag{1.3.20}$$

where $d\Omega = g(x)\, dx\, d\xi$, $l = 1, 2, \dots, N$.

Equalities (1.3.20) represent a system of linear equations for α_{Ni}, where $i = 1, 2, \dots, N$.

We now prove that under the hypotheses of Theorem 1.3.2 the system (1.3.20) has a unique solution $\alpha_N = (\alpha_{Ni})_{i=1}^N$ for any $F \in H^2(\Omega)$. It suffices to prove that the homogeneous version of (1.3.20) has only trivial solution. Assume the contrary. Let the homogeneous version of (1.3.20) have a nonzero solution $\bar{\alpha}_N = (\bar{\alpha}_{Ni})$. Setting $F = 0$ in (1.3.20), replace α_N by $\bar{\alpha}_N$, multiply the lth equation by $-2\bar{\alpha}_{Nl}$ and sum from 1 to N with respect to l. As a result, we obtain

$$-2\int_\Omega g^{ij}\nabla'_j \frac{\partial}{\partial \xi^i}(L\bar{u}_N - F)\bar{u}_N \, d\Omega = 0, \tag{1.3.21}$$

where $\bar{u}_N = \sum_{i=1}^{N} \bar{\alpha}_{Ni}\omega_i$. Since

$$-2g^{ij}\nabla'_j \frac{\partial}{\partial \xi^i}(L\bar{u}_N)\bar{u}_N = -2\nabla'_j\Big(g^{ij}\frac{\partial}{\partial \xi^i}(L\bar{u}_N)\bar{u}_N\Big) + 2g^{ij}\frac{\partial}{\partial \xi^j}(L\bar{u}_N)\nabla'_j\bar{u}_N,$$

by virtue of equalities (1.3.12), (1.3.16), and the condition $u_N = 0$ on Γ_1, from (1.3.21) we get

$$J(\bar{u}_N) \equiv \int_\Omega \left(g^{ij} \nabla'_j \bar{u}_N \nabla'_i \bar{u}_N - R^{si}_{rl} \xi^r \xi^l \frac{\partial \bar{u}_N}{\partial \xi^s} \frac{\partial \bar{u}_N}{\partial \xi^i} \right) \mathrm{d}\Omega$$

$$+ \int_{\partial\Omega} (Lu_N)^2 \frac{g\, \mathrm{d}S}{\left[\sum\limits_{i=1}^n (g_{ij}\xi^j)^2 \right]^{1/2}} = 0. \qquad (1.3.22)$$

From the assumptions of Theorem 1.3.2 it follows that $\nabla'_j \bar{u}_N = 0$ for $j = 1, 2, \dots, n$, i.e., $L\bar{u}_N = 0$, which yields $\bar{u}_N = 0$ in Ω_δ since $\bar{u}_N = 0$ on Γ_1, $\bar{u}_N \in C^3(\Omega)$, and $L \equiv \mathrm{d}/\mathrm{d}s$. The system ω_i is linearly independent and therefore we have $\alpha_{Ni} = 0$ $(i = 1, 2, \dots, n)$, which contradicts the assumption that $\alpha_N \neq 0$. The contradiction shows that the homogeneous version of system (1.3.20) has only trivial solution. Thus, system (1.3.20) has a unique solution $\alpha_N = (\alpha_{Ni})$ for any function $F \in H^2(\Omega_\delta)$.

We now estimate u_N in terms of F. For this purpose, we multiply the lth equation of (1.3.20) by $-2\alpha_{Nl}$ and sum from 1 to N with respect to l. As a result, we get

$$-2 \int_\Omega g^{ij} \nabla'_j \frac{\partial}{\partial \xi^i} (Lu_N) u_N \, \mathrm{d}\Omega = -2 \int_\Omega g^{ij} \nabla'_j \frac{\partial F}{\partial \xi^i} u_N \, \mathrm{d}\Omega. \qquad (1.3.23)$$

As follows from (1.3.16), and since $u_N = 0$ on Γ_1, the right-hand side of (1.3.23) can be estimated as follows:

$$2 \left| \int_\Omega u_N g^{ij} \nabla'_j \frac{\partial F}{\partial \xi^i} \, \mathrm{d}\Omega \right| \leq \frac{2}{\delta_0} \int_\Omega |\nabla_\xi F|^2 \, \mathrm{d}\Omega + \frac{\delta_0}{2} \int_\Omega \sum_{j=1}^n |\nabla'_j u_N|^2 \, \mathrm{d}\Omega,$$

where $\delta_0 > 0$ is such that for any $\eta = (\eta_1, \dots, \eta_n) \in \mathbb{R}^n$

$$g^{ij} \eta_i \eta_j \geq \delta_0 |\eta|^2, \quad |\nabla_\xi F|^2 = \sum_{i=1}^n F^2_{\xi^i}.$$

The left-hand side of (1.3.23) is equal to $J(u_N)$ (see (1.3.22)). Thus, from (1.3.23) it follows that

$$J(u_N) \leq \frac{2}{\delta_0} \int_\Omega |\nabla_\xi F|^2 \, \mathrm{d}\Omega + \frac{\delta_0}{2} \int_\Omega \sum_{i=1}^n |\nabla'_i u_N|^2 \, \mathrm{d}\Omega.$$

By virtue of (1.3.3), the foregoing inequality yields

$$\int_\Omega \sum_{i=1}^n |\nabla_i' u_N|^2 \, d\Omega \le C, \quad \int_{\Gamma_2} (Lu_N)^2 \, ds \le C,$$

which, in its turn, implies

$$\int_\Omega (Lu_N)^2 \, d\Omega \le C.$$

Using the inequalities above and the regularity condition for a family of geodesics in $\mathcal{M}$, we prove estimates of the form

$$\int_\Omega u_N^2 \, d\Omega \le C, \qquad \int_{\Gamma_2} u_N^2 \, ds \le C,$$

where C, generally, stands for different constants that are independent of N.

Consider the domain

$$\widetilde{\Omega} = \{(x', \xi', s) \,:\, x' \in \partial D, \; g_{ij}(x')\xi'^i \xi'^j < 1, \; \xi' \ne 0, \; s \in (0, l(x', \xi'))\},$$

where $l(x', \xi')$ is the length of the part of the geodesic $\gamma(x', \xi')$ lying within D with initial point $x' \in \partial D$ and direction ξ' at this point. Take a geodesic $\gamma(x^0, \xi^0)$ with initial point $(x_0, \xi_0) \in \Omega \setminus \{(x, \xi) : \xi = 0\}$ (i.e., consider the solution to the Cauchy problem (1.3.4) – (1.3.5)). The geodesic intersects ∂D at a point x', and its direction at this point is ξ'. Associate the point $(x_0, \xi_0) \in \Omega \setminus \{(x, \xi) : \xi = 0\}$ with the point $(x', \xi', s) \in \widetilde{\Omega}$, where s is the length of the part of the geodesic $\gamma(x^0, \xi^0)$ with initial point x^0 and direction ξ^0 at x^0 whose endpoints are (x^0, ξ^0) and the point of intersection of the geodesic and ∂D. Now, conversely, choose a geodesic $\gamma(x', \xi')$ with initial point (x', ξ'). Take its part of length s, with endpoints $x' \in \partial D$ and $x^0 \in D$, and take the tangent vector ν of $\gamma(x', \xi')$ at x^0. Let $\xi^0 = |\xi'|\nu$, where $|\xi'| = g_{ij}(x')\xi'^i \xi'^j$ (the length of the tangent vector of a geodesic is constant; see Dubrovin, Novikov, and Fomenko, 1979), that is, we have associated the point $(x', \xi', s) \in \widetilde{\Omega}$ with $(x^0, \xi^0) \in \Omega \setminus \{(x, \xi) : \xi = 0\}$. The uniqueness of the solution of the Cauchy problem (1.3.4) – (1.3.5) implies that we have thus constructed a one-to-one correspondence: $x = x(x', \xi', s)$, $\xi = \xi(x', \xi', s)$.

According to the regularity condition for a family of geodesics, the following inequality holds in $\overline{\Omega}$:

$$0 < \left| \frac{\partial(x', \xi', s)}{\partial(x, \xi)} \right| \le n_1. \tag{1.3.24}$$

Set $\tilde{u}_N(x', \xi', s) = u_N(x, \xi)$. Since $L = \dfrac{\mathrm{d}}{\mathrm{d}s}$ on $\gamma(x, \xi)$ and $u_N(x, \xi) = 0$ on Γ_1, we have

$$\tilde{u}_N(x', \xi', s) = \int_0^s Lu(x(t), \xi(t))\,\mathrm{d}t,$$

where $x(t) = x(x', \xi', t)$ and $\xi(t) = \xi(x', \xi', t)$. Therefore

$$|\tilde{u}_N|^2 = C \int_0^{l(x',\xi')} \left[L\tilde{u}_N(x', \xi', t)\right]^2 \mathrm{d}t, \tag{1.3.25}$$

or, as follows from (1.3.24),

$$\int_{\Gamma_1} |\tilde{u}_N(x', \xi', s)|^2\,\mathrm{d}\Gamma_1$$

$$\leq C_1 \int_{\Gamma_1} \left(\int_0^{l(x',\xi')} (Lu_N)^2\,\mathrm{d}t \right) \leq C_1 \int_\Omega (Lu_N)^2\,\mathrm{d}\Omega \leq C,$$

where $\mathrm{d}\Gamma_1 = g(x')\,\mathrm{d}s\,\mathrm{d}\xi$ and $\mathrm{d}s$ is the surface element of ∂D.

Since D is bounded, using (1.3.24), from the foregoing inequality we easily obtain

$$\int_\Omega u_N^2\,\mathrm{d}\Omega \leq C_1 \int_\Omega (Lu_N)^2\,\mathrm{d}\Omega.$$

The last inequality shows that the summand $\int_\Omega u^2\,\mathrm{d}\Omega$ does not affect the equivalence of norms of the space $H(\nabla')$. One can prove in a similar way that

$$\int_{\Gamma_2} u_N^2\,\mathrm{d}s \leq C.$$

Thus, the sets $\{Lu_N\}$ and $\{u_N\}$ are bounded in $L_2(\Omega)$ and the sets of traces of these functions on Γ are bounded in $L_2(\Gamma_2)$. Since $L_2(\Omega)$ is a Hilbert space, the sets $\{Lu_N\}$ and $\{u_N\}$ are weakly compact in $L_2(\Omega)$. Consequently, there exist subsequences (we again denote them by $\{Lu_N\}$ and $\{u_N\}$) such that $u_N \overset{w}{\to} u$ and $Lu_N \overset{w}{\to} \widetilde{\lambda}$ in $L_2(\Omega)$ as $N \to \infty$, where the symbol $\overset{w}{\to}$ means weak convergence. Since the operator L is weakly closed in $L_2(\Omega)$, we have $\widetilde{\lambda} = Lu$.

Similarly, without loss of generality, we may assume that $u_N\big|_{\Gamma_2} \overset{w}{\to} u^{(2)}$ and $Lu_N\big|_{\Gamma_2} \overset{w}{\to} \lambda^{(2)}$ in $L_2(\Gamma_2)$ as $N \to \infty$, where $u_N\big|_{\Gamma_2}$ and $Lu_N\big|_{\Gamma_2}$ are the traces of the functions u_N and Lu_N on Γ_2. We now prove that $\lambda^{(2)} = Lu^{(2)}$. It is easy to verify that

$$|\mathbf{n}_\xi| = \left[\sum_1^n \mathbf{n}_{\xi^i}^2 \right]^{1/2} \neq 0$$

on the surface Γ_2. Let $(x_0, \xi_0) \in \Gamma_2$ and $\mathbf{n}_{\xi^{i_0}} \neq 0$. For the sake of simplicity, assume that $i_0 = n$. Then there exists a neighbourhood $U(x_0, \xi_0)$ of the point (x_0, ξ_0) in which the Jacobian of the transformation $(x, \xi) \to (z, \eta)$ of the form

$$z^i = x^i,\ \eta^l = z^l,\ \eta^n = g_{ij}\xi^i\xi^j, \quad i, j = 1, 2, \dots, n,\ l = 1, 2, \dots, n-1,$$

is nonzero. Using the equality $\Gamma_{i,kj} + \Gamma_{k,ij} = \partial g_{ik}/\partial x^j$, one can verify directly that $LF_2 = 0$, where $F_2 = g_{ij}\xi^i\xi^j - 1$. Therefore, the derivative with respect to η^n (the "outward" derivative) does not appear in the expression for Lu in terms of the variables (z, η) on the surface

$$\Gamma_2(x_0, \xi_0) = \{(z, \eta) : (z, \eta \in \{\eta_n = 0\} \cap U(x_0, \xi_0)\},$$

which means that the operator Lu is "interior" on the surface Γ_2. Since $u_N\big|_{\Gamma_2}$ and $Lu_N\big|_{\Gamma_2}$ weakly converge in $L_2(\Gamma_2)$ to $u^{(2)}$ and $\lambda^{(2)}$, respectively, it follows that the same is true in $L_2(\Gamma_2(x_0, \xi_0))$. The differential operator L is "interior" and weakly closed in $L_2(\Gamma_2(x_0, \xi_0))$, which implies $\lambda^{(2)} = Lu^{(2)}$ on $\Gamma_2(x_0, \xi_0)$. Covering Γ_2 with pieces of the form $L_2(\Gamma_2(x_0, \xi_0))$, we obtain $\lambda^{(2)} = Lu^{(2)}$ on Γ_2.

Using inequality (1.3.25), which is true for functions $z \in C^1(\Omega)$, it is easy to prove that u has a trace on Γ_1 that belongs to $L_2(\Omega)$ if $u, Lu \in L_2(\Omega)$. Since $u_N = 0$ on Γ_1, $u = 0$ on Γ_1.

Below we show that the weak limit $u(x, \xi)$ of the sequence $\{u_N\}$ in $L_2(\Omega)$ as $N \to \infty$ is one of the desired components of the solution to the problem (1.3.18) - (1.3.19).

In (1.3.20), transferring the operator $g^{ij}\nabla'_j\, \partial/\partial\xi^i$ to $\beta\omega_l(x, \xi)$ and taking into account (1.3.16) and the fact that $u_N = 0$ on Γ_1, for $N \geq l$ we have

$$\int_\Omega (Lu_N - F)\, g^{ij}\frac{\partial}{\partial \xi^i}\nabla'_j(\omega_l)\, \mathrm{d}\Omega - \int_{\Gamma_2} g\, g^{ij}(Lu_N - F)\nabla'_j(\omega_l)\, \mathbf{n}_{\xi^i}\, \mathrm{d}s = 0.$$

Since the linear span of the set of functions $\omega_l(x, \xi)$ is everywhere dense in $\overset{\circ}{H}{}^{-}_{1,2}(\Omega)$, passing to the limit as $N \to \infty$ in the last equality we get

$$\int_\Omega (Lu - F) g^{ij}\nabla'_j\frac{\partial\eta}{\partial\xi^i}\, \mathrm{d}\Omega - \int_{\Gamma_2} g^{ij}(Lu - F)\nabla'_j\eta\, \mathbf{n}_{\xi^i}\, g\, \mathrm{d}s = 0$$

for any $\eta \in \overset{\circ}{H}{}^{-}_{1,2}(\Omega)$.

We put $\lambda = Lu - F$. The final part of the proof of Theorem 1.3.2 is similar to that of Theorem 1.1.2, Section 1.1. □

The existence theorem for Problem 1.3.2 analogous to Theorem 1.3.2 can be proved in exactly the same way.

Consider Problem 1.3.2 (or Problem 1.3.3) for Ω_δ with $\delta = 1$. Then the solution $u(x,\xi)$ of equation (1.3.1) (which holds in Ω_0) is known on $\Gamma_{11} = \{(x,\xi) : (x,\xi) \in \Gamma_1, \, g_{ij}\xi^i\xi^j = 1\}$, i.e., we have $u|_{\Gamma_{11}} = u_0$. Suppose that $u_0 \in H^2(\Gamma_{11})$ and $g_{ij} \in C^{[\frac{n+1}{2}]+3}$. Because $\partial D \in C^{[n/2]+4}$, there exists a function $\tilde{u}_0 \in H^{\frac{n+5}{2}}(\Gamma_1)$ such that $\tilde{u}_0 = u_0$ on Γ_{11} (see Lions and Magenes, 1972). Thus, we have Problem 1.3.2 in the case where $\delta = 0$. Consequently, we can formulate the existence theorem analogous to Theorem 1.3.2 for Problem 1.3.3 under the hypotheses of Theorem 1.3.2 and the additional assumption that $g_{ij} \in C^{[\frac{n+1}{2}]+3}$ and $g_{ij}\xi^i\xi^j = 1$.

1.4. A PLANAR PROBLEM OF INTEGRAL GEOMETRY

The problem considered in this section is formulated as follows (see Mukhometov, 1975). Let D be a bounded and simply connected planar domain with smooth boundary Γ:

$$\begin{gathered} x = \xi(z), \quad y = y(z), \qquad z \in (0, L) \\ \xi(0) = \xi(L), \quad \eta(0) = \eta(L), \end{gathered} \tag{1.4.1}$$

where z is the length of the curve Γ.

We define smooth curves in $\overline{D}$ by the equations

$$x = \varphi(x_0, y_0, \theta, s), \quad y = \psi(x_0, y_0, \theta, s), \tag{1.4.2}$$

where (x_0, y_0) is the initial point of a curve and θ is its tangent angle at this point. The variable parameter s is the arc length. The domain of definition of the functions φ and ψ is

$$T = \{(x_0, y_0, \theta, s) \,:\, (x_0, y_0) \in D, \, \theta \in (0, 2\pi), \, s \in (0, l(x_0, y_0, \theta))\},$$

where $l(x_0, y_0, \theta)$ is defined as follows: for the curve with initial point (x_0, y_0) and tangent angle θ at this point, $l(x_0, y_0, \theta)$ is the length of the part of this curve with one endpoint at (x_0, y_0) and the other one being the intersection of the curve and the boundary of the domain. The conditions specified below will provide that the function $l(x, y, \theta)$ be always well defined.

Consider the set of curves defined by (1.4.2) as a two-parameter family of curves $K(\gamma, z)$ satisfying the following conditions:

(a) for any two points of $\overline{D}$ there is a unique curve in $K(\gamma, z)$ that passes through these points; every curve in $K(\gamma, z)$ intersects Γ only at two points $(\xi(z), \eta(z))$ and $(\xi(\gamma), \eta(\gamma))$; the lengths of the curves in $K(\gamma, z)$ are uniformly bounded;

(b) $\varphi \in C^3(T)$, $\psi \in C^3(T)$, and all the derivatives of φ and ψ are uniformly bounded in T;

(c) $\dfrac{1}{s}\dfrac{D(\varphi, \psi)}{D(\theta, s)} \geq C > 0$, where C is a constant;

(d) $\varphi(x, y, 0, s) = \varphi(x, y, 2\pi, s)$, $\psi(x, y, 0, s) = \psi(x, y, 2\pi, s)$, and the same is true for the derivatives of φ and ψ of up to the third order.

From the definition of the set of curves (1.4.2) and the set T it follows that for any point in D there is a unique curve in $K(\gamma, z)$ passing through this point in a given direction.

We now formulate the main problem. Let

$$v(\gamma, z) = \int_{K(\gamma,z)} \lambda(x, y)\, ds \tag{1.4.3}$$

Given the set of curves $K(\gamma, z)$ and the function $v(\gamma, z)$ defined in the rectangle $0 < \gamma,\ z < L$, determine the function $\lambda(x, y)$ on $\overline{D}$. In other words, we know the function v defined for any pair of points of the boundary Γ. For every such pair, there is a unique curve in K that passes through these points, and the meaning of the function v is given by (1.4.3).

We shall analyze a problem equivalent to the one formulated above with a certain differential equation instead of equation (1.4.3). We introduce the function

$$u(x, y, z) = \int_{K(x,y,z)} \lambda\, ds, \tag{1.4.4}$$

where $K(x, y, z)$ is the part of a curve from $K(\gamma, z)$ that connects the points (x, y) and $(\xi(z), \eta(z))$, and $(x, y, z) \in \Omega = \overline{D} \times (0, L)$. The function $u(x, y, z)$ has the following differential properties:

1. $u(x, y, z) \in C(\Omega)$ has continuous derivatives up to the second order on $\Omega_1 = \Omega \setminus \{(\xi(z), \eta(z), z) : z \in (0, L)\}$;

2. the derivatives u_x, u_y, and u_z are bounded in Ω_1, and u_{xz}, u_{yz}, u_{xy} may have a singularity of the type $\left[(x - \xi(z))^2 + (y - \eta(z))^2\right]^{-1/2}$.

These properties follow from conditions (a) – (d) and the fact that $\lambda(x,y) \in C^2(\overline{D})$. In addition, if λ belongs to the class of finite functions (with D as a support), then $u_{xz}, u_{yz}, u_{xy} \in C(\Omega)$ (see Mukhometov, 1975).

Differentiating (1.4.4) in the direction of the line tangent to $K(x,y,z)$ at the point (x,y) with angle $\theta(x,y,z)$, we obtain

$$L_1 u \equiv \frac{\partial u}{\partial x}\cos\theta + \frac{\partial u}{\partial y}\sin\theta = \lambda(x,y) \tag{1.4.5}$$

The function $\theta(x,y,z)$ is differentiable in Ω_1. The derivative θ_z has singularities of the type $\left[(x-\xi(z))^2 + (y-\eta(z))^2\right]^{-1/2}$.

The data of the problem for equation (1.4.3) provide the following boundary condition for $u(x,y,z)$:

$$u\big|_{\partial D\times(0,L)} = u_0, \tag{1.4.6}$$

and the condition of periodicity in z for $(x,y,z) \in \Omega$:

$$u(x,y,z) = u(x,y,z+L). \tag{1.4.7}$$

Consider a problem which is more general than (1.4.5) – (1.4.7): assume that λ depends also on z and satisfies the condition

$$\int_\Omega \lambda \frac{\partial}{\partial z}\left(\frac{\partial\varphi}{\partial y}\cos\theta - \frac{\partial\varphi}{\partial x}\sin\theta\right) \mathrm{d}x\,\mathrm{d}y\,\mathrm{d}z = 0 \tag{1.4.8}$$

for any $\varphi \in C^0_L(D) = \{\varphi : \varphi = 0 \text{ on } \partial D\times(0,L),\ \varphi \in C^3_L\}$, where C^3_L is the set of real-valued functions $\varphi \in C^3(\Omega)$ that are L-periodic in z.

Condition (1.4.8) implies that λ satisfies the equation

$$\lambda_{zx}\sin\theta - \lambda_{zy}\cos\theta + \lambda_z\theta_x\cos\theta + \lambda_z\theta_y\sin\theta = 0$$

in the generalized sense.

Problem 1.4.1. Find a pair of functions (u,λ) that satisfy equation (1.4.5) and conditions (1.4.6) – (1.4.8).

A scalar product in $C^0_L(\Omega)$ is introduced by the formula

$$(u,z)_\theta = \int_\Omega (u_x z_x + u_y z_y)\theta\,\mathrm{d}x\,\mathrm{d}y\,\mathrm{d}z.$$

We also introduce the following designations:

$$Au \equiv (\sin\theta(L_1 u)_z)_x - ((L_1 u)_z\cos\theta)_y, \tag{1.4.9}$$

$H_\theta(\Omega)$ is the completion of $C_L^0(\Omega)$ with respect to the norm $\|u\|_\theta^2 = (u,u)_\theta$, $\Gamma''(A)$ is the set of all functions $u(x,\varphi) \in H_\theta(\Omega)$ for any of which there exists $y \in L_2(\Omega)$ such that for all $\eta \in C_L^0(\Omega)$

$$\int_\Omega L_1 u(\eta_x \sin\theta - \eta_y \cos\theta)_z \,dx\,dy\,dz = \int_\Omega y\,\eta \,dx\,dy\,dz,$$

and $y = Au$.

$\Gamma(A)$ and $\Gamma'(A)$ are defined as in Section 1.1, with the set $C^3_{\pi 0}$ replaced by $C_L^0(\Omega)$, and the differential expression A replaced by the expression given by (1.4.9); weak convergence $u_k \to u$ is assumed to be in the space $H_\theta(\Omega)$; $H_m^L(\Omega)$ is the completion of $C_L^3(\Omega)$ with respect to the norm of H_m, where $m = 1, 2$.

Now we are ready to formulate an assertion for the following problem equivalent to Problem 1.4.1.

Problem 1.4.1′. Find a pair of functions (u,λ) defined in Ω satisfying the equation

$$L_1 u = \lambda + F, \tag{1.4.5′}$$

provided that F is given, the trace of the solution of (1.4.5′) on $\partial D \times (0, L)$ is zero, λ satisfies condition (1.4.8) and $u(x,y,z)$ satisfies (1.4.7).

Theorem 1.4.1. *Let conditions (a) – (d) hold for the family of curves (1.4.2), and let $F \in H_2^L(\Omega)$. Then there exists a unique solution (u,λ) of Problem 1.4.1′ such that $u \in \Gamma(A)$, $u \in H_\theta(\Omega)$, $\lambda \in L_2(\Omega)$, and*

$$\|u\|_{H_\theta(\Omega)} + \|\lambda\| \le C\,\|F\|_{H_1^L(\Omega)},$$

where C depends on the size of the domain D.

The proof is similar to that of Theorems 1.1.1 and 1.1.2.

We give an example of a family of curves which satisfies the assumptions of Theorem 1.1.1, but does not satisfy the assumptions of Theorem 1.4.1.

A family of curves $\{\Gamma\}$ in the ball $B = \{(x,y) : x^2+y^2 < 1/2\}$ is defined as follows. Let $y(x,x_0,y_0,y_1)$ be the solution to the Cauchy problem

$$\begin{cases} y'' = x - yy', \\ y(x_0) = y_0, \quad y'(x_0) = y_1, \end{cases} \tag{1.4.10}$$

where $x_0^2 + y_0^2 < 1/2$. Let $\Gamma(x_0,y_0,y_1)$ denote the part of the integral curve $(x, y(x,x_0,y_0,y_1))$ of the Cauchy problem (1.4.10) that lies inside B, and set

$$\{\Gamma\} = \{\Gamma(x_0,y_0,y_1) : (x_0,y_0) \in B,\ y_1 \in \mathbb{R}^1\}.$$

Then the curvature of $\Gamma(x_0, y_0, y_1)$ at the point (x, y) is easily calculated: $f(x, y, \varphi) = (x\cos\varphi - y\sin\varphi)\cos^2\varphi$, where $y' = \tan\varphi$. Since y depends on (x_0, y_0, y_1), f implicitly depends on the same arguments. One can easily verify that f satisfies the assumptions of Theorem 1.1.1 in the ball B.

We now prove that any two distinct points of B are connected by at least two curves in $\{\Gamma\}$. Indeed, let $\Gamma(x_0, y_1^0, y_1^1)$ be an integral curve of the Cauchy problem

$$\begin{cases} y_1'' = x - y_1 y_1', \\ y_1(x_0) = y_1^0, \;\; y_1'(x_0) = y_1^1, \end{cases}$$

and let $\Gamma(x_0, y_1^0, y_2^1)$ be an integral curve of the Cauchy problem

$$\begin{cases} y_2'' = -y_2 y_2', \\ y_2(x_0) = y_1^0, \;\; y_2'(x_0) = y_2^1. \end{cases}$$

The curves $\Gamma(x_0, y_1^0, y_1^1)$ and $\Gamma(x_0, y_1^0, y_2^1)$ pass through the point (x_0, y_1^0).

We shall show that these curves intersect at some other point of B for y_1^1 and y_2^1 sufficiently large; consequently, the point (x_0, y_1^0) and the point of intersection of $\Gamma(x_0, y_1^0, y_1^1)$ and $\Gamma(x_0, y_1^0, y_2^1)$ belong to two distinct curves, which means that the set of curves $\{\Gamma\}$ specified above does not satisfy the assumptions of Theorem 1.4.1.

To this end, we consider the difference $\tilde{y} = y_2 - y_1$. Then $\tilde{y}(x)$ satisfies the following relations:

$$\begin{cases} \tilde{y} = y_1\tilde{y}' + y_2'\tilde{y}, \\ \tilde{y}(x_0) = 0, \;\; \tilde{y}'(x_0) = y_2^1 - y_1^1. \end{cases} \tag{1.4.11}$$

Using the well-known change of variables (see Petrovskii, 1964)

$$\tilde{y}(x) = z_1(x)\exp\Big(\frac{1}{2}\int_{x_0}^{x} y_1(\xi)\,\mathrm{d}\xi\Big),$$

the equation in (1.4.11) is transformed to the form

$$z_1'' + \beta(x)z_1 = 0,$$

where $\beta(x) = -y_1^2/4 - y_1'/2 - y_2'$.

Now we consider the equation

$$z_2'' + \lambda^2 z_2 = 0, \tag{1.4.12}$$

where λ is sufficiently large, so that the function $\tilde{z}_2 = \dfrac{\sin \lambda x}{\lambda}$, which satisfies equation (1.4.12), has several zeros in a neighbourhood $U(x_0) \subset B$ of x_0. Choose y_1 and y_2^1 so that $\beta^2 > \lambda^2$ in some neighbourhood $U'(x_0) \subset U(x_0)$ of x_0. It is possible because $y_i = \tan \varphi_i$ $(i = 1, 2)$, φ varies in the interval from 0 to 2π, and y_i are continuously differentiable in B.

Then, by Sturm's Theorem on the zeros of solutions of linear homogeneous second order equations (see Petrovskii, 1964), z_1 has zeros in $U'(x_0)$, and therefore the same is true for $\tilde{y}(x)$. This fact means that the curves $\Gamma(x_0, y_1^0, y_1^1)$ and $\Gamma(x_0, y_1^0, y_2^1)$ intersect in B at least twice.

1.5. CERTAIN PROBLEMS OF TOMOGRAPHY

In the conclusion of Chapter 1 we give examples of certain inverse problems which are of great importance in applications, namely, in tomography. The analysis of these problems is reduced to the study of the problems considered above. Tomography is a method designed to obtain images of internal components of opaque bodies.

Let $u(x, \nu)$ be the particle radiation flux. Then propagation of the flux in a medium with linear attenuation constant a and scattering coefficient b is described by the transport equation (see Anikonov, 1985; Tikhonov, Arsenin, and Timonov, 1987):

$$\nu \cdot \nabla_x u + au = b \int_{S^2} u(x, \nu')\, d\nu', \tag{1.5.1}$$

where $x \in D$, $D \subset \mathbb{R}^3$ is a bounded domain with boundary $\partial D \in C^3$, and $S^2 = \{\nu : \nu \in \mathbb{R}^3, |\nu| = 1\}$.

The boundary condition for (1.5.1) has the form

$$u(x, \nu) = h(x, \nu), \quad (x, \nu) \in \partial D \times S^2, \quad \langle \mathbf{n}_x, \nu \rangle < 0, \tag{1.5.2}$$

where $\mathbf{n}_x$ is the unit vector of the outward normal to the surface ∂D at the point $x \in \partial D$. The function $h(x, \nu)$ represents the intensity of external radiation.

It is known (see Anikonov, 1985) that by choosing an external source of a special type, namely, the so-called "slit source", the scattering effect can be made small enough, so that (1.5.1) could be replaced by the equation

$$\nu \cdot \nabla_x u + au = 0. \tag{1.5.3}$$

From the results of Vladimirov (1961) it follows that for $\beta < (4\pi d)^{-1}$ (where d is the diameter of D) there exists a unique solution of equation (1.5.1) that satisfies condition (1.5.2) if $a \in C^2(\overline{D})$, $b \in C^1(\overline{D})$, $|b| < \beta$, $\beta > 0$, and D is a strictly convex domain.

Consider the following inverse problem.

Problem 1.5.1. Find a pair of functions (u, a) satisfying equation (1.5.3), condition (1.5.2), and the condition

$$u(x,\nu) = H(x,\nu), \quad (x,\nu) \in \partial D \times S^2, \qquad \langle \mathbf{n}_x, \nu\rangle \geq 0. \tag{1.5.4}$$

We set $\Gamma_1 = \partial D \times S^2$ and

$$u_0 = \begin{cases} h(x,\nu) & \text{if } \langle \mathbf{n}_x, \nu\rangle < 0, \\ H(x,\nu) & \text{if } \langle \mathbf{n}_x, \nu\rangle \geq 0. \end{cases}$$

Theorem 1.5.1. *Suppose that $u_0 \in H^3(\Gamma_1)$, $u_0 \neq 0$. Then there exists a unique solution to Problem 1.5.1 such that $u \in \Gamma(A) \cap H_1(\Omega)$, $a \in L_2(\Omega)$, $u \neq 0$ in $\Omega = D \times S^2$, where $\Gamma(A)$ is the set introduced in Section 1.2 (with $F_1 = F_2 = 0$), and a satisfies (1.2.2).*

Proof. Since $u \neq 0$ in Ω, we divide equation (1.5.3) by $u(x,\nu)$ and denote $\ln u$ by w. Then we arrive at Problem 1.2.1 for w with the data $w_0 = \ln u_0$ ($F_1 = F_2 = 0$). Consequently, the assertion of Theorem 1.5.1 follows from Theorems 1.2.1 and 1.2.2 (see Remark 1.2.2). □

Set

$$\lambda = b \int_{S^2} u(x,\nu')\,d\nu'.$$

Problem 1.5.2. Find a pair of functions (u, b) that satisfy (1.5.1), provided that $a = 0$, λ satisfies (1.2.2), and u is known on Γ_1: $u\big|_{\Gamma_1} = u_0$.

Theorem 1.5.2. *Let $u_0 \in H^3(\Gamma_1)$, $u_0 \neq 0$. Then there exists a unique solution (u, λ) of Problem 1.5.2 such that $(u, \lambda) \in \Gamma(A) \cap H_1(\Omega) \times L_2(\Omega)$ and λ satisfies (1.2.2).*

The assertion of Theorem 1.5.2 follows from Theorems 1.2.1 and 1.2.2. If the obtained solution is such that

$$\int_{S^2} u(x,\nu')\,d\nu' \neq 0,$$

then

$$b = \lambda\Big(\int_{S^2} u(x,\nu')\,\mathrm{d}\nu'\Big)^{-1}.$$

Remark 1.5.1. If $\nu \cdot \nabla_x u$ in (1.5.1) and (1.5.3) is replaced by Lu from Sections 1.1 – 1.4 and 2.2, then the theorems analogous to Theorems 1.5.1 and 1.5.2 hold for the corresponding versions of Problems 1.5.1 and 1.5.2.

Chapter 2.

Inverse problems for kinetic equations

The main subject of Chapter 2 is the uniqueness of solutions to inverse problems for kinetic equations; the existence of solutions is also proved for certain problems (Sections 2.2 and 2.4). Physical interpretation of such problems consists in finding forces of particle interaction, scattering indicatrices, radiation sources, and other physical parameters. Some of these problems are connected with integral geometry (Sections 2.1, 2.4, and 2.7).

2.1. THE PROBLEM OF INTEGRAL GEOMETRY AND AN INVERSE PROBLEM FOR THE KINETIC EQUATION

Let $D \in \mathbb{R}^n$ be a domain of variables x that lies in the half-space $x_1 > 0$ and has piecewise smooth boundary. Introduce the following designations: $V_\delta = \{v \in R^n : v_1 > v_1^0 > 0,\ |'v| < \delta\}$, $D_T = \{(x,t) : x \in D, t \in (-T,0)\}$.

Consider the Cauchy problem for the system

$$\ddot{z} = f(\tau, z, \dot{z}) \tag{2.1.1}$$

with the data

$$\begin{aligned} z(\tau)\big|_{\tau=t} &= x, \\ \dot{z}(\tau)\big|_{\tau=t} &= v = \tan\varphi, \end{aligned} \tag{2.1.2}$$

where $z = (z_1, z_2, \dots, z_n)$, $\varphi = (\varphi_1, \varphi_2, \dots, \varphi_n)$, $\dot{z}_i = dz_i/ d\tau$, and φ_i is the angle between the axis $O\tau$ and the line tangent to the curve $z_i(\tau)$ at the point $\tau = t$.

For a vector function $f = (f_1, f_2, \dots, f_n)$ we assume that $f \in C(D_T \times V_\delta)$ and $f_1 > 0$. Let $\lambda(x,t) \in C_0(D_T)$ be the unknown function and $z(x, v, t)$ be the solution to the Cauchy problem (2.1.1) – (2.1.2). We introduce a function $u(x, v, t)$ as follows:

$$u(x,v,t) = \int_{z(x,v,t)} \lambda(x(\tau), \tau)\, d\tau, \tag{2.1.3}$$

where $(x, t) \in D_T$ and $v \in V_\delta$.

Differentiating equation (2.1.3) along the direction $v = \tan\varphi$ (i.e., with respect to t), we get

$$u_t + \langle v, \nabla_x u\rangle + \langle f, \nabla_v u\rangle = \lambda(x,t). \tag{2.1.4}$$

Assume that $u(x, v, t)$ is defined for $v \in V_\delta$, $t \in (-T, 0)$, $'x \in {}'D$, for $x_1 = 0$, $D = (0,1) \times {}'D$.

The last condition means that we know the integrals of $\lambda(x,t)$ over all integral curves of the Cauchy problem (2.1.1) – (2.1.2) that pass through the plane $x_1 = 0$ touching it at tangent angle φ, where $\tan\varphi = v \in V_\delta$.

We have $f_1 > 0$, $v_1 > v_1^0$, and the curves $z(x, v, t)$ lie "above" the plane $z_1 = \tau$ increasing monotonically as $v_1 \to \infty$. Therefore, $u(x, v, t) \to 0$ as $v_1 \to \infty$, and condition $\lambda \in C_0(D_T)$ and equality (2.1.3) imply that $u = 0$ when $t = 0$ and $x_1 = 1$.

Thus, we arrive at the following problem.

Problem 2.1.1. Find a pair of functions satisfying equation (2.1.4) defined in the domain $\Omega = D_T \times V_\delta$, provided that u, $u_{v_1} \in L_2(v_1^0, \infty)$ for all $(x,t) \in D_T$, $|'v| < \delta$, and the following conditions hold:

$$u\big|_{x_1=0} = u_1, \qquad u\big|_{x_1=1} = u\big|_{t=0} = 0. \tag{2.1.5}$$

Throughout what follows we shall assume that the function λ in equation (2.1.4) depends on all variables except v_1: $\lambda = \lambda(x, 'v, t) \in C^1(\Omega)$.

Theorem 2.1.1. *Assume that $f_1 > 0$, f does not depend on x_1, $'f$ does not depend on v_1, and $f_1, f_{1v_1} \in L_2(v_1^0, \infty)$. Then Problem 2.1.1 has at most one solution (u, λ) such that $u \in C^2(\Omega)$, $\nabla u_{v_1} \in L_2(v_1^0, \infty)$.*

Proof. Consider the homogeneous version of Problem 2.1.1. Extend $u(x,v,t)$ so that $u = 0$ for $x_1 \geq 1$ and $x_1 \leq 0$, and apply the Fourier transform with respect to x_1 to equation (2.1.4). We have

$$\hat{u}_t + v_1(\mathrm{i}\,\xi)\hat{u} + \langle 'v,\ \nabla_{'x}\hat{u}\rangle + \langle f,\ \nabla_v\hat{u}\rangle = \hat{\lambda}(\xi, 'x, 'v, t).$$

Differentiating this equation with respect to v_1 and putting $\hat{u} = r + \mathrm{i}z$, we obtain

$$\begin{cases} z_{tv_1} - v_1 z_{v_1}\xi + \langle 'v,\ \nabla_{'x} r_{v_1}\rangle + \langle f,\ \nabla_v r_{v_1}\rangle - \xi z + f_{1v_1} r_{v_1} = 0, \\ z_{tv_1} + v_1 r_{v_1}\xi + \langle 'v,\ \nabla_{'x} z_{v_1}\rangle + \langle f,\ \nabla_v z_{v_1}\rangle + \xi r + f_{1v_1} z_{v_1} = 0. \end{cases} \tag{2.1.6}$$

From (2.1.5) it follows that

$$r\big|_{t=0} = 0, \quad z\big|_{t=0} = 0. \tag{2.1.7}$$

Since $u(x,v,t)$ is finite with respect to x_1, the function $\hat{u}$ is entire with respect to ξ for fixed $('x,v,t) \in \Omega$. Therefore, to find $\hat{u}$ we have to know its value together with the values of its derivatives at $\xi = 0$. When $\xi = 0$, the system (2.1.6) turns into one equation

$$s_{tv_1} + \langle 'v,\ \nabla_{'x} s_{v_1}\rangle + \langle f,\ \nabla_v s_{v_1}\rangle + f_{1v_1} s_{v_1} = 0. \tag{2.1.8}$$

From (2.1.7) we obtain

$$s_{v_1}\big|_{t=0} = 0. \tag{2.1.9}$$

Let numbers M and M_1 be such that

$$\forall x \in D \ \ |x_i| \leq M, \quad \forall (x,v,t) \in \Omega \ \ |f_i| \leq M_1, \qquad i = 1, 2, \dots, n,$$

and let

$$'\Omega_{\alpha,\beta} = \{('x,v,t) \,:\, \beta t + \alpha > |'x - 'x^0| + |'v|^2,\ v_1 > v_1^0\},$$

where $\alpha > 0$ and $\beta > \delta(n-2)(M + M_1) + 1$.

We now multiply equation (2.1.8) by the quantity $-2s_{v_1}\exp(k\varkappa)$, where $\varkappa = \beta t - \frac{1}{2}(|'v|^2 + |'x - 'x^0|^2)$, $k > 0$. We get

$$\begin{aligned} 0 &= -2s_{v_1}\exp(k\varkappa)\big(s_{tv_1} + \langle 'v, \nabla_{'x} s_{v_1}\rangle + f_{1v_1} s_{v_1} + \langle f, \nabla_v s_{v_1}\rangle\big) \\ &= k s_{v_1}^2 \exp(k\varkappa)\Big[\beta - \frac{1}{k} f_{1v_1} - \sum_{i=2}^{n}\Big((x_i - x_i^0)v_i + v_i f_i - \frac{1}{k} f_{iv_i}\Big)\Big] \\ &\quad + d_1(s) + d_2(s), \end{aligned} \tag{2.1.10}$$

where

$$d_1(s) = -\sum_{i=2}^{n} \left(v_i \exp(k\varkappa) s_{v_1}^2\right)_{x_i} - (s_{v_1} \exp(k\varkappa))_t$$

$$d_2(s) = -\sum_{i=1}^{n} \left(s_{v_i} f_i \exp(k\varkappa)\right)_{v_i}$$

Integrate equality (2.1.10) over $'\Omega_{\alpha,\beta}$ and pass to the limit as $k \to \infty$. From the conditions $\beta > \delta(n-2)(M+M_1)+1$, $f_1 > 0$, and since $s_{v_1} \to 0$ as $v_1 \to \infty$, we obtain

$$\int_{'\Omega_{\alpha,\beta}} s_{v_1}^2 \exp(k\varkappa)\,\mathrm{d}'\Omega_{\alpha,\beta} \le \int_{'\Omega_{\alpha,\beta}} (d_1(s) + d_2(s))\,\mathrm{d}'\Omega_{\alpha,\beta}$$

Since $s \to 0$ as $v_1 \to 0$ and $s_{v_1} = 0$ when $t = 0$, from the foregoing inequality it follows that $s = 0$ on $'\Omega_{\alpha,\beta}$. Consequently, (2.1.6) and (2.1.7) imply that $r = z = 0$ on $'\Omega_{\alpha,\beta}$.

Differentiating the system (2.1.6) with respect to ξ and then setting $\xi = 0$, we arrive at the problem (2.1.8) – (2.1.9) for the function s_ξ. Thus, we may assume that $s_\xi = 0$ in $'\Omega_{\alpha,\beta}$. Similar arguments yield

$$\left.\frac{\partial^m s}{\partial \xi^m}\right|_{\xi=0} = 0, \qquad m = 2, 3, \dots .$$

Recalling that s is an entire function with respect to ξ, we have $s \equiv 0$ on $(-\infty,\infty) \times {}'\Omega_{\alpha,\beta}$, i.e., $\tilde{u} \equiv 0$ on $(-\infty,\infty) \times {}'\Omega_{\alpha,\beta}$, which gives $u = 0$ in $\Omega_{\alpha,\beta}$, where $\Omega_{\alpha,\beta} = \{(x,v,t) : x_1 \in (0,1),\ ('x,v,t) \in {}'\Omega_{\alpha,\beta}\}$.

Next, varying $'x^0 = (0, x_2^0, x_3^0, \dots, x_n^0)$ in $'D = \{x \in D : x_1 = 0\}$, we establish that $u = 0$ in the domain of the form

$$\Omega_\beta = \{(x,v,t) : (x,v,t) \in \Omega,\ 0 > t > -\alpha/\beta\}.$$

Repeating the same arguments for the domain of the form $\Omega^\beta = \{(x,v,t) : (x,v,t) \in \Omega,\ t < -\alpha/\beta\}$, we prove that $u = 0$ in $\Omega \cap \{0 > t > -2\alpha/\beta\}$, and so on. As a result, we have $u = 0$ in Ω, and from (2.1.4) it follows that $\lambda = 0$ in Ω. This completes the proof of Theorem 2.1.1. □

2.2. LINEAR KINETIC EQUATION

We analyze the problem of finding a solution and the right-hand side of the kinetic equation in the case where the values of the solution are known on the boundary of a domain. The first part of this section is devoted to the uniqueness of solutions of the inverse problem in question, whereas the question of existence is analyzed in the second part.

Let Ω be a domain in the Euclidean space $\mathbb{R}^{2n}$, $n \geq 1$. For the variables $(x, v) \in \Omega$ we assume $x \in D$, $v \in G$, where D and G are domains in $\mathbb{R}^n$ with boundaries of class C^2; $\partial\Omega$ is the boundary of Ω, $\partial\Omega = \overline{\Gamma}_1 \cup \overline{\Gamma}_2$, where $\Gamma_1 = \partial D \times G$, $\Gamma_2 = D \times \partial G$. By $H_{1,C}(\Omega)$ we denote the set of real-valued functions $u(x, v) \in L_2(\Omega)$ having generalized derivatives u_{x_i}, u_{v_i}, $u_{x_i v_i}$ which belong to $L_2(\Omega)$, where $i = 1, 2, \dots, n$; $\overset{\circ}{H}_{1,C} = \overset{\circ}{H}_1 \cap H_{1,C}$.

Consider the following linear kinetic equation in Ω:

$$Lu \equiv \{u, H\} \equiv \sum_{i=1}^{n} \left(\frac{\partial H}{\partial v_i} \frac{\partial u}{\partial x_i} - \frac{\partial H}{\partial x_i} \frac{\partial u}{\partial v_i} \right) = \lambda(x, v) \tag{2.2.1}$$

with the right-hand side λ such that

$$\langle \lambda, \widehat{L}\eta \rangle = 0, \tag{2.2.2}$$

for any $\eta \in H_{1,2}(\Omega)$ whose trace on $\partial\Omega$ is zero. Here

$$\widehat{L} = \sum_{i=1}^{n} \frac{\partial^2}{\partial x_i \partial v_i},$$

$\{u, H\}$ is the Poisson bracket of u and H, and $H_{1,2}(\Omega)$ is the set defined in Section 1.3.

It is easy to check that condition (2.2.2) holds, for example, for any function λ of the form $\lambda = g(x) + \psi(v)$, where g and ψ are continuously differentiable functions.

Equation (2.2.1) is extensively used in plasma physics and astrophysics (see Alexeev, 1982; Liboff, 1979). In applications, u represents the number (or the mass) of particles in the unit volume element of the phase space in the neighbourhood of the point (x, v), $\nabla_x H$ is the force acting on a particle, λ is the collisional term characterizing the variation of u caused by particle collisions.

We define $\widetilde{C}_0^3 = \{\varphi : \varphi \in C^3(\Omega),\, \varphi = 0 \text{ on } \partial\Omega\}$ and select a subset $\{w_1, w_2, \dots\}$ of $\widetilde{C}_0^3$ which is orthonormal and everywhere dense in $L_2(\Omega)$. Let $\mathcal{P}_n$ be the orthogonal projector of $L_2(\Omega)$ onto $\mathfrak{M}_n$, where $\mathfrak{M}_n$ is the linear span of $\{w_1, w_2, \dots\}$. By $\Gamma(A)$ we denote the set of functions u with the following properties:

1. $u \in L_2(\Omega)$, $Au \in L_2(\Omega)$ in the generalized sense, where $Au = \widehat{L}Lu$;
2. there exists a sequence $\{u_k\} \subset \widetilde{C}_0^3$ such that $u_k \to u$ in $L_2(\Omega)$ and $\langle Au_k,\, u_k\rangle \to \langle Au,\, u\rangle$ as $k \to \infty$.

The condition that $Au \in L_2(\Omega)$ in the generalized sense means that there exists a function $f \in L_2(\Omega)$ such that $\langle u,\, A^*\varphi\rangle = -\langle f,\, \varphi\rangle$ and $Au = f$ for all $\varphi \in C_0^\infty(\Omega)$, where A^* is the differential operator conjugate to A in the sense of Lagrange. It is easy to prove that

$$H_3 \cap H_1^0 \subset \Gamma''(A) \cap \overset{\circ}{H}_{1,C} \subset \Gamma(A) \subset L_2(\Omega),$$

where $\Gamma''(A)$ is the set of functions $u \in L_2(\Omega)$ such that $Au \in L_2(\Omega)$ in the generalized sense.

We turn to the study of the following inverse problem (see Anikonov and Amirov, 1983):

Problem 2.2.1. Determine a pair of functions (u, λ) defined in Ω from equation (2.2.1), the given Hamiltonian $H(x, v) \in C^2(\Omega)$, and the trace of the solution u on the boundary $\partial\Omega$, i.e.,

$$u\big|_{\partial\Omega} = u_0. \tag{2.2.3}$$

1. We now formulate the uniqueness theorem for Problem 2.2.1.

Theorem 2.2.1. *Assume that $H \in C^2(\overline{\Omega})$ and the following inequalities hold for all $\xi \in \mathbb{R}^n$, $(x, v) \in \overline{\Omega}$:*

$$\sum_{i,j=1}^{n} \frac{\partial^2 H}{\partial v_i v_j}\xi^i\xi^j \ge \alpha|\xi|^2, \qquad \sum_{i,j=1}^{n} \frac{\partial^2 H}{\partial x_i \partial x_j}\xi^i\xi^j \le 0, \tag{2.2.4}$$

where α is a positive number. Then Problem 2.2.1 has at most one solution (u, λ) such that $u \in \Gamma(A)$ and $\lambda \in L_2(\Omega)$.

Proof. Let (u, λ) be a solution to Problem 2.2.1 such that $u = 0$ on $\partial\Omega$ and $u \in \Gamma(A)$. Then (2.2.1) and (2.2.2) imply $Au = 0$. Since $u \in \Gamma(A)$, there exists a sequence $\{u_k\}$ such that $u_k \in \widetilde{C}_0^3$ and, as $k \to \infty$,

$$\begin{aligned} &u_k \to u \quad \text{in } L_2, \\ &\langle Au_k, u_k\rangle \to 0. \end{aligned} \tag{2.2.5}$$

Observing that $u_k = 0$ on $\partial\Omega$, we get

$$-\langle Au_k, u_k\rangle = \sum_{i=1}^{n}\Big\langle \frac{\partial}{\partial v_i}(Lu_k),\, u_{kx_i}\Big\rangle. \tag{2.2.6}$$

To transform the right-hand side of (2.2.6), we consider the identity

$$\begin{aligned} &\sum_{j=1}^{n}\frac{\partial u_k}{\partial x_j}\frac{\partial}{\partial v_j}(Lu_k) \\ &\quad= \frac{1}{2}\sum_{i,j=1}^{n}\Big(\frac{\partial^2 H}{\partial v_i\partial v_j}\frac{\partial u_k}{\partial x_i}\frac{\partial u_k}{\partial x_j} - \frac{\partial^2 H}{\partial x_i\partial x_j}\frac{\partial u_k}{\partial v_i}\frac{\partial u_k}{\partial v_j}\Big) \\ &\qquad+ \frac{1}{2}\sum_{i,j=1}^{n}\frac{\partial}{\partial v_j}\Big[\frac{\partial u_k}{\partial x_j}\Big(\frac{\partial u_k}{\partial x_i}\frac{\partial H}{\partial v_i} - \frac{\partial u_k}{\partial v_i}\frac{\partial H}{\partial x_i}\Big)\Big] \\ &\qquad- \frac{1}{2}\sum_{i,j=1}^{n}\frac{\partial}{\partial x_j}\Big[\frac{\partial u_k}{\partial v_j}\Big(\frac{\partial u_k}{\partial x_i}\frac{\partial H}{\partial v_i} - \frac{\partial H}{\partial x_i}\frac{\partial u_k}{\partial v_i}\Big)\Big] \\ &\qquad+ \frac{1}{2}\sum_{i,j=1}^{n}\frac{\partial}{\partial x_i}\Big(\frac{\partial H}{\partial v_i}\frac{\partial u_k}{\partial x_j}\frac{\partial u_k}{\partial v_j}\Big) - \frac{1}{2}\sum_{i,j=1}^{n}\frac{\partial}{\partial v_i}\Big(\frac{\partial H}{\partial x_i}\frac{\partial u_k}{\partial x_j}\frac{\partial u_k}{\partial v_j}\Big). \end{aligned} \tag{2.2.7}$$

Taking into account the geometry of the domain Ω and the condition that $u_k = 0$ in Ω, from (2.2.7) we obtain

$$-\langle Au_k, u_k\rangle = J(u_k), \tag{2.2.8}$$

where

$$J(u) \equiv \frac{1}{2}\sum_{i,j=1}^{n}\int_\Omega\Big(\frac{\partial^2 H}{\partial v_i\partial v_j}\frac{\partial u}{\partial x_i}\frac{\partial u}{\partial x_j} - \frac{\partial^2 H}{\partial x_i\partial x_j}\frac{\partial u}{\partial v_i}\frac{\partial u}{\partial v_j}\Big)\,d\Omega.$$

Since Ω is bounded and $u_k = 0$ on $\partial\Omega$, by virtue of (2.2.4) we have

$$J(u_k) \geq \frac{1}{2}\int_\Omega\sum_{i,j=1}^{n}\frac{\partial^2 H}{\partial v_i\partial v_j}\frac{\partial u_k}{\partial x_i}\frac{\partial u_k}{\partial x_j}\,d\Omega \geq C\int_\Omega u_k^2\,d\Omega, \tag{2.2.9}$$

where $C > 0$ does not depend on k. Thus, from (2.2.5), (2.2.8), and (2.2.9) it follows that

$$C \int_{\Omega} u^2 \, d\Omega \leq 0,$$

i.e., $u = 0$ in Ω. Then (2.2.1) implies $\lambda(x, v) = 0$. This completes the proof of Theorem 2.2.1. □

Remark 2.2.1. The example constructed in Remark 1.3.2 shows that the geometry of the domain Ω is essential for the uniqueness of the solution to Problem 2.2.1 in Theorem 2.2.1.

Note that the assertion of Theorem 2.2.1 still holds for unbounded domains under certain conditions for the decreasing of the solution together with its first order derivatives at infinity (see Amirov, 1986d). To prove the theorem in this case, Ω should be "approximated" by bounded domains.

The following example shows that condition (2.2.4) is essential for the assertion of Theorem 2.2.1 to hold.

Suppose $\Omega = (-\infty, \infty) \times (-\infty, \infty)$ and $H = (x^2 + v^2)/2$ (evidently, the second inequality in (2.2.4) does not hold). It easy to verify that the functions $u = \exp\{-r(x^2 + v^2)\}$ and $\lambda = 0$ satisfy equation (2.2.1) with $H = (x^2 + v^2)/2$ ($r > 0$ is an arbitrary fixed number and $u(x, v)$ decreases sufficiently rapidly at infinity together with its derivatives).

We note also that Problem 2.2.1 will have infinitely many solutions if condition (2.2.2) is not imposed on λ, which is evident from the following example. Suppose that some function $w \in C^2(\overline{\Omega})$ coincides with the given function u_0 on ∂D, where $u_0 = 0$ on Γ_2. Such functions exist, which will be proved in the second part of this section (see Remark 2.2.2). Then the pair (w, Lw) is a solution to Problem 2.2.1.

2. In this part of the section, the existence theorem for Problem 2.2.1 will be proved using the Galerkin method.

Problem 2.2.1′. Given the equation

$$Lu = \lambda + F, \tag{2.2.10}$$

where λ satisfies (2.2.2) and F is a known function in $H_2(\Omega)$, find the pair of functions (u, λ) under the condition that

$$u\big|_{\partial\Omega} = 0. \tag{2.2.11}$$

Theorem 2.2.2. *Assume $H \in C^2(\overline{\Omega})$ and the following inequalities hold for all $(x, v) \in \overline{\Omega}$, $\xi \in \mathbb{R}^n$:*

$$\sum_{i,j=1}^{n} \frac{\partial^2 H}{\partial v_i \partial v_j} \xi^i \xi^j \geq \alpha_1 |\xi|^2, \qquad \sum_{i,j=1}^{n} \frac{\partial^2 H}{\partial x_i \partial x_j} \xi^i \xi^j \leq -\alpha_2 |\xi|^2, \tag{2.2.4$'$}$$

where α_1 and α_2 are positive numbers and $F \in H_2(\Omega)$. Then there exists a solution (u, λ) of Problem 2.2.1$'$ such that $u \in \Gamma(A)$, $u \in H_1(\Omega)$, $\lambda \in L_2(\Omega)$.

Proof. Analyze the following auxiliary problem: Solve the equation

$$Au = \mathcal{F}, \tag{2.2.12}$$

under the condition that

$$u\big|_{\partial\Omega} = 0, \tag{2.2.13}$$

where $\mathcal{F} = \widehat{L}F$.

This problem will be referred to as the problem (2.2.12), (2.2.13).

We shall seek an approximate solution to problem (2.2.12), (2.2.13) in the form

$$u_N = \sum_{i=1}^{N} \alpha_{Ni} w_i$$

to satisfy the following relations:

$$\langle Au_N - \mathcal{F}, w_i \rangle = 0, \qquad i = 1, 2, \dots, N, \tag{2.2.14}$$

or

$$\Big\langle \sum_{j=1}^{N} \alpha_{Ni} Aw_j - \mathcal{F}, w_i \Big\rangle = 0, \qquad i = 1, 2, \dots, N, \tag{2.2.15}$$

where α_{Ni} are the unknown numbers.

Equalities (2.2.15) constitute a system of linear algebraic equations for α_{Ni}, $i = 1, 2, \dots, N$.

We shall prove that system (2.2.15) has a unique solution $\alpha_N = (\alpha_{Ni})_{i=1}^{N}$ under the hypotheses of Theorem 2.2.2. To this end, we multiply the ith equation of the system by $-2\alpha_{Ni}$ and sum from 1 to N with respect to i. We obtain

$$-2\langle Au_N, u_N \rangle = 0.$$

By (2.2.8) we have

$$-2\langle Au_N, u_N\rangle = 2J(u_N) = 0. \tag{2.2.16}$$

The assumptions of Theorem 2.2.2 imply $\nabla u_N = 0$. Hence $u_N = 0$ in Ω as a result of the conditions $u_N = 0$ on Γ_1 and $u \in \widetilde{C}_0^3(\Omega)$. Since the system $\{w_i\}$ is linearly independent, we get $\alpha_{Ni} = 0$, $i = 1, 2 \ldots, N$. Thus, the homogeneous version of system (2.2.15) has only trivial solution.

Therefore, system (2.2.15) has a unique solution $\alpha_N = (\alpha_{Ni})_{i=1}^N$ for any function $F \in H_2(\Omega)$.

We now estimate u_N in terms of F. For this purpose, we multiply the ith equation of system (2.2.15) by $-2\alpha_{Ni}$ and sum from 1 to N with respect to i. Since $\mathcal{F} = \widehat{L}F$, we obtain

$$-2\langle Au_N, u_N\rangle = -2\langle \widehat{L}F, u_N\rangle. \tag{2.2.17}$$

With the observation that $u_N = 0$ on $\partial\Omega$, the right-hand side of (2.2.17) is estimated as follows:

$$\begin{aligned} -2\langle \widehat{L}F, u_N\rangle &= 2\int_\Omega \sum_{i=1}^n \frac{\partial F}{\partial v_i}\frac{\partial u_N}{\partial x_i}\,d\Omega \\ &\le \beta\int_\Omega |\nabla_v F|^2\,d\Omega + \beta^{-1}\int_\Omega |\nabla_x u_N|^2\,d\Omega, \end{aligned}$$

where $\beta^{-1} < \alpha_1$.

As was shown above, the left-hand side of (2.2.17) is equal to $2J(u_N)$ (see (2.2.8)). Then from (2.2.17) we obtain

$$2J(u_N) \le \beta\int_\Omega |\nabla_v F|^2\,d\Omega + \beta^{-1}\int_\Omega |\nabla_x u_N|^2\,d\Omega.$$

Recalling that Ω is bounded and $u_N = 0$ on Γ_1, the last inequality implies

$$\|u_N\|_{\overset{\circ}{H}_1(\Omega)} \le C\,\|\,|\nabla_v F|\,\|, \tag{2.2.18}$$

where the constant $C > 0$ does not depend on N.

As follows from inequality (2.2.18), the final part of the proof is exactly the same as that of Theorem 1.1.2 with the use of estimate (1.1.16). □

Remark 2.2.2. For example, if $\partial D \in C^{[\frac{n+1}{2}]+3}$, $u_0 \in H^{\frac{n+1}{2}+2}(\partial\Omega)$, and $u_0 = 0$ on Γ_2, then Problem 2.2.1 can be reduced to Problem 2.2.1′. Indeed, under these conditions there exists a function $w \in H^{[n/2]+3}(\Omega)$ such that $w = 0$ on Γ_2. For any fixed "section" $\widetilde{\Gamma}_1 = \partial D \times v_0$ of the surface Γ_1 we can construct a function $w(x, v_0)$ such that $w(x, v_0) \in H^{[n/2]+3}(D)$, $w(x, v_0) = u_0(x, v_0)$ on $\widetilde{\Gamma}_1$, and $\|w\|_{H^{[n/2]+3}} \le C \|u_0\|_{H^{(n+5)/2}}$ (see Theorem 8.3, Lions and Magenes (1972); Theorem 2 of Section 4, Chapter 3, Mikhailov (1976)). Since $u_0 \in H^{(n+5)/2}$, it is obvious that $w \in H^{n/2+3}(\Omega)$ and $w = 0$ on Γ_2. We set $u_1 = u - w$. The function u_1, taken for u, satisfies the equation $Lu = \lambda - F$ with $F = Lw$ and the condition $u\big|_{\partial\Omega} = 0$.

Remark 2.2.3. From the proofs of Theorems 2.2.1 and 2.2.2 it is clear that similar results can be obtained for the analogous problem if we replace equation (2.2.1) by a more general kinetic equation with terms representing absorption and scattering, i.e.,

$$Lu + \beta u + \int_G K(x, v, v')u(x, v')\, dv' = \lambda(x, v) + F(x, v),$$

where the given functions β, K, and F are sufficiently smooth. In this case, α, α_1, and α_2 in Theorems 2.2.1 and 2.2.2 will depend also on β and K.

2.3. A MODIFICATION OF PROBLEM 2.2.1

1. Consider two equations in the domain Ω that was introduced in Section 2.2:

$$\{u, H\} = \lambda_1(x, v), \tag{2.3.1}$$

$$\sum_{i=1}^{n} \Big(\frac{\partial H}{\partial v_i} a_i - \frac{\partial H}{\partial x_i} b_i \Big) = \lambda_2(x, v), \tag{2.3.2}$$

where $a = (a_1, \ldots, a_n) \equiv \nabla_x g$, $b = (a_1, \ldots, a_n) \equiv \nabla_v \overline{G}$, $g_{x_i v_i} = \overline{G}_{x_i v_i}$, $i = 1, 2, , \ldots, n$; g, $\overline{G} \in C^3(\overline{\Omega})$, $\lambda_k \in C^2$, $k = 1, 2$; $H, u \in C^2(\overline{\Omega})$, and λ_k satisfy the equation

$$\sum_{j=1}^{n} \frac{\partial^2 \lambda_k}{\partial x_j \partial v_j} = 0, \qquad k = 1, 2. \tag{2.3.3}$$

Assume that

$$u_{x_i} = a_i\big|_{\partial\Omega}, \quad u_{v_i} = b_i\big|_{\partial\Omega}, \quad \lambda_1 = \lambda_2\big|_{\Gamma_1}. \tag{2.3.4}$$

Problem 2.3.1. Find out if relations (2.3.4) are valid also in Ω.

Theorem 2.3.1. *Assume that (2.2.4′) holds for H, $\partial H/\partial v_i$ are independent of x, and $\partial H/\partial x_i$ are independent of v. Moreover, assume that relations (2.3.1) – (2.3.4) hold. Then $a = \nabla_x u$, $b = \nabla_v u$, and $\lambda_1 = \lambda_2$ in Ω.*

Proof. Subtract (2.3.1) from (2.3.2) and put $\lambda = \lambda_2 - \lambda_1$. We have

$$\sum_{i=1}^{n}\left[\frac{\partial H}{\partial v_i}(u_{x_i} - a_i) - \frac{\partial H}{\partial x_i}(u_{v_i} - b_i)\right] = \lambda. \tag{2.3.5}$$

Apply the operator

$$L \equiv \sum_{i=1}^{n}(u_{x_i} - a_i)\frac{\partial}{\partial v_i}$$

to both sides of (2.3.5). Evidently, λ satisfies (2.3.3). Thus, (2.3.5) implies

$$\begin{aligned} 2J &\equiv 2\sum_{j=1}^{n}(u_{x_j} - a_j)\frac{\partial}{\partial v_j}\left\{\sum_{i=1}^{n}\left[\frac{\partial H}{\partial v_i}(u_{x_i} - a_i) - \frac{\partial H}{\partial x_i}(u_{v_i} - b_i)\right]\right\} \\ &= 2\sum_{j=1}^{n}(u_{x_j} - a_j)\frac{\partial}{\partial v_j}\lambda = 2\sum_{j=1}^{n}\frac{\partial}{\partial x_j}\left[(u - g)\frac{\partial \lambda}{\partial v_j}\right]. \end{aligned} \tag{2.3.6}$$

On the other hand,

$$\begin{aligned} J = \sum_{i,j=1}^{n}&\left[\frac{\partial^2 H}{\partial v_i \partial v_j}(u_{x_i} - a_i)(u_{x_j} - a_j) - \frac{\partial^2 H}{\partial x_i \partial x_j}(u_{v_i} - b_i)(u_{v_j} - b_j)\right] \\ + \sum_{i,j=1}^{n}&\left\{\left[\frac{\partial H}{\partial v_i}(u_{x_i} - a_i)(u_{x_j} - a_j)\right]_{v_j} - \left[\frac{\partial H}{\partial v_i}(u - g)_{v_i}(u_{x_i} - a_i)\right]_{x_j}\right. \\ &+ \left[(u_{x_j} - a_j)(u - g)_{v_j}\frac{\partial H}{\partial v_i}\right]_{x_i} - \left[\frac{\partial H}{\partial x_i}(u_{x_j} - a_j)(u_{v_i} - b_i)\right]_{v_j} \\ &+ \left[\frac{\partial H}{\partial x_i}(u_{v_i} - b_i)(u - \overline{G})_{x_j}\right]_{v_j} + \left[\frac{\partial H}{\partial x_i}(u_{v_i} - b_i)(u_{v_j} - b_j)\right]_{x_j} \\ &\left. - \left[\frac{\partial H}{\partial x_i}(u - \overline{G})_{x_j}(u_{v_j} - b_j)\right]_{v_i}\right\}. \end{aligned}$$

Integrating both sides of (2.3.6) over Ω and using (2.3.4), we obtain

$$\int_\Omega \sum_{i,j=1}^{n} \Big[\frac{\partial^2 H}{\partial v_i \partial v_j}(u_{x_i}-a_i)(u_{x_j}-a_j) - \frac{\partial^2 H}{\partial x_i \partial x_j}(u_{v_i}-b_i)(u_{v_j}-b_j)\Big]\, d\Omega = 0$$

Hence, by virtue of (2.2.4′) we have $u_{x_i} = a_i$, $u_{v_i} = b_i$ in Ω $(i = 1, 2, \dots, n)$. Therefore, (2.3.5) implies that $\lambda = 0$ in Ω, which completes the proof. □

2. Consider the kinetic equation

$$\{u, H\} = \lambda(x, v) \tag{2.3.7}$$

in the domain Ω, where the right-hand side satisfies equation (2.3.3).

Problem 2.3.2. Determine the functions $u(x,v)$ and $\lambda(x,v)$ from the given Hamiltonian $H \in C^2(\overline{\Omega})$, the conditions

$$\nabla_x u\big|_{\partial\Omega} = u_0, \qquad \lambda\big|_{\Gamma_1} = \lambda_0, \tag{2.3.8}$$

and the value of $u(x,v)$ at a point $(x^0, v^0) \in \Omega$: $u(x^0, v^0) = u^1$.

Theorem 2.3.2. *If inequalities (2.2.4′) hold in Ω, then Inverse Problem 2.3.2 has at most one solution (u, λ) such that $u \in C^2(\overline{\Omega})$ and $\lambda \in C^2(\overline{\Omega})$.*

Proof. Suppose that the pair $u(x,v)$, $\lambda(x,v)$ is a solution of Inverse Problem 2.3.2 such that $u^0 = \lambda^0 = u^1 = 0$. We show that $u = \lambda = 0$ in Ω. Since λ satisfies equation (2.3.3), we have

$$2\sum_{j=1}^{n} \frac{\partial u}{\partial x_j}\Big[\frac{\partial}{\partial v_j}\Big(\sum_{i=1}^{n} \frac{\partial u}{\partial x_i}\frac{\partial H}{\partial v_i} - \frac{\partial u}{\partial v_i}\frac{\partial H}{\partial x_i}\Big)\Big] = 2\sum_{j=1}^{n} \frac{\partial}{\partial x_j}\Big(u\frac{\partial \lambda}{\partial v_j}\Big).$$

By virtue of identity (2.2.7) and relation (2.3.8) with $u^0 = \lambda_0 = 0$, we get

$$\sum_{i,j=1}^{n} \int_\Omega \Big(\frac{\partial^2 H}{\partial v_i \partial v_j} u_{x_i} u_{x_j} - \frac{\partial^2 H}{\partial x_i \partial x_j} u_{v_i} u_{v_j}\Big)\, d\Omega = 0.$$

Using (2.2.4), from the last equality we obtain $u_{x_i} = u_{v_i} = 0$, $i = 1, 2, \dots, n$, and equation (2.3.7) implies that $\lambda = 0$ in Ω. Since $u(x_0, v_0) = u^1 = 0$, it follows that $u \equiv 0$ in Ω, and the proof is complete. □

Note that the vector function u^0 in (2.3.8) is not arbitrary. It is the gradient of a certain function with respect to x and, consequently, the components of u^0 are related by certain differential formulas.

Problem 2.3.3. Find the pair of functions (u, λ) defined in Ω that satisfy equation (2.3.7) and the conditions

$$\nabla u\big|_{\partial\Omega} = u_0, \qquad \nabla_v \lambda\big|_{\partial\Omega} = \lambda_0, \qquad u(x_0, v_0) = u_1,$$

where (x_0, v_0) is a point in Ω.

Theorem 2.3.3. *Let inequalities (2.2.4′) hold. Then Problem 2.3.3 has at most one solution* (u, λ) *such that* $u \in C^2(\overline{\Omega})$, $\lambda \in C^2(\overline{\Omega})$.

The proof is similar to that of Theorem 2.3.2.

It should be noted that the geometry of the domain Ω is not essential for the assertion of Theorem 2.3.3 to hold if $\nabla_v \lambda$ is given on the entire boundary $\partial\Omega$. However, the geometry of the domain Ω is known to be essential for the existence and uniqueness of solutions of inverse problems for the kinetic equation (see Remark 2.2.1).

2.4. ONE-DIMENSIONAL KINETIC EQUATION

In this section we analyze three inverse problems for one-dimensional kinetic equation.

Problem 2.4.1. Given the stationary one-dimensional kinetic equation

$$lu \equiv vu_x + fu_v = \lambda(x)g(x, v) + F(x, v), \tag{2.4.1}$$

in the domain $\Omega = \{(x, v) : x \in (0, 1),\ v \in (-1, 1)\}$, where f, g, F are given functions, find $u(x, v)$ and $\lambda(x)$, provided that

$$u(0, v) = 0, \tag{2.4.2}$$

$$u(x, 1) = u(x, -1) = 0. \tag{2.4.3}$$

From the results formulated in Section 2.2 it follows that if λ (one of the unknown functions in Problem 2.2.1) depends on (x, v), then the following conditions are necessary for Problem 2.2.1 to be well-posed:

1. $u(x, v)$ is known on $\partial\Omega$;
2. λ satisfies the equality $\lambda_{xv} = 0$.

It turns out that if λ does not depend on v, then conditions (2.4.2) and (2.4.3) in Problem 2.4.1 are sufficient for finding (u, λ).

We shall first establish the uniqueness of the solution to Problem 2.4.1, and then prove the existence theorem for the corresponding Hilbert space choosing a special "basis" by means of the Galerkin method.

Consider the system of functions $\{x^k\}$, $k = 1, 2, 3, \dots$, $x \in (0,1)$. It is well known that, supplemented by 1, this system becomes a "basis" in $L_2(0,1)$ and $C[0,1]$, i.e., the set of finite linear combinations of the obtained system is everywhere dense in $L_2(0,1)$ and $C[0,1]$.

Let $\mathcal{M}$ be the set of all finite linear combinations of functions of the form $x^k\omega(v)$, $k = 1, 2, \dots$, where $\omega(v)$ are twice continuously differentiable real-valued functions such that $\omega(-1) = \omega(1) = 0$.

By the symbols $\overset{\circ}{H}_1(\Omega)$ and $\overset{\circ}{H}_{1,2}(\Omega)$ we denote the Hilbert spaces that are the completions of $\mathcal{M}$ with respect to the norms of the spaces H_1 and $H_{1,2}$, correspondingly (see the definition of the space $H_{1,2}$ in Section 1.3).

For the sake of simplicity we set $g \equiv 1$ in Ω in equation (2.4.1). The case $g \neq 0$ in Ω is analyzed similarly.

Remark 2.4.1. The homogeneous boundary conditions (2.4.2) and (2.4.3) in Problem 2.4.1 can be replaced by the conditions

$$u(0, v) = u_1(v),$$
$$u(x, -1) = u_2(x), \quad u(x, 1) = u_3(x),$$

with the following concordance conditions: $u_1(1) = u_1(-1) = u_2(0) = u_3(0) = 0$, where u_i, $i = 1, 2, 3$, are any given continuous functions. Then the function $U = u - \varphi$, where $\varphi = 0.5(1+v)u_3(x) + 0.5(1-v)u_2(x) + u_1(v)$, satisfies equation (2.4.1) with free term $F + l\varphi$ and homogeneous boundary conditions.

Theorem 2.4.1. *Let a function f be continuously differentiable in $\overline{\Omega}$, and let the following conditions hold:*

1. $f_x > 0$,
2. $f(1, v) \leq 0$.

Then Problem 2.4.1 has at most one solution (u, λ) such that $u \in \overset{\circ}{H}_{1,2}(\Omega)$ and $\lambda \in L_2(0,1)$.

Proof. Suppose that (u, λ) is a solution to the homogeneous version of Problem 2.4.1 ($F = 0$, $g \equiv 1$) such that $u \in \overset{\circ}{H}_{1,2}(\Omega)$ and $\lambda \in L_2(0,1)$. Then, differentiating (2.4.1) with respect to v (in the generalized sense), we have

$$vu_{xv} + fu_{vv} + u_x + f_v u_v = 0.$$

Let $u_n \in \mathcal{M}$ and $u_n \to u$ as $n \to \infty$ with respect to the norm of $\overset{\circ}{H}_{1,2}(\Omega)$. By the definition of $\overset{\circ}{H}_{1,2}(\Omega)$ and since $f \in C^1(\overline{\Omega})$,

$$2\int_\Omega u_{nx}\frac{\partial}{\partial v}(lu_n)\,d\Omega \to 2\int_\Omega u_x \frac{\partial}{\partial v}(lu)\,d\Omega \qquad \text{as } n \to \infty.$$

Consequently, as follows from the identity

$$2u_{nx}\frac{\partial}{\partial v}(lu_n) = u_{nx}^2 + f_x u_{nv}^2 + (u_{nx}^2 v)_v + 2(fu_{nv}u_{nx})_v - (fu_{nv}^2)_x \tag{2.4.4}$$

and the conditions $\frac{\partial}{\partial v}(lu) = 0$, $u_n \in \mathcal{M}$, and $f(1,v) \le 0$, we have

$$\begin{aligned} 0 &= 2\int_\Omega u_x\frac{\partial}{\partial v}(lu)\,d\Omega = \lim_{n\to\infty} 2\int_\Omega u_{nx}\frac{\partial}{\partial v}(lu_n)\,d\Omega \\ &= \lim_{n\to\infty}\Big[\int_\Omega (u_{nx}^2 + f_x u_{nv}^2)\,d\Omega - \int_{-1}^{1} f(1,v)u_{nv}^2\,dv\Big] \\ &\ge \lim_{n\to\infty}\int_\Omega (u_{nx}^2 + f_x u_{nv}^2)\,d\Omega = \int_\Omega (u_x^2 + f_x u_v^2)\,d\Omega, \end{aligned}$$

which yields $u = 0$ by virtue of (2.4.2) and (2.4.3), whereas equation (2.4.1) with $F = 0$ and $g = 1$ gives $\lambda = 0$. □

Theorem 2.4.2. *Under the hypotheses of Theorem 2.4.1, suppose that $F \in C^1(\Omega)$. Then Problem 2.4.1 has a solution (u, λ) such that $u \in \overset{\circ}{H}_1(\Omega)$ and $\lambda \in L_2(0,1)$.*

Proof. Consider the following problem:
Solve the equation

$$\frac{\partial}{\partial v}(lu) = F_v(x,v), \tag{2.4.5}$$

under the conditions (2.4.2), (2.4.3).

This problem will be called the problem (2.4.5), (2.4.2), (2.4.3). We shall seek its "approximate" solution in the form

$$u_n(x,v) = \sum_{i=1}^{n} u_{ni}(v)x^i,$$

where the functions u_{ni} are determined from the conditions

$$\Big\langle \frac{\partial}{\partial v}(lu_n),\, x^j \Big\rangle_{L_2(0,1)} = \langle F_v,\, x^j\rangle_{L_2(0,1)} \qquad j = 0,1,\dots,n-1, \quad (2.4.6)$$

$$u_{ni}(-1) = u_{ni}(1) = 0 \qquad\qquad i = 1,2,\dots,n. \qquad (2.4.7)$$

Equalities (2.4.6) represent a linear system of ordinary second order differential equations for $u_{ni}(v)$. Since the determinant of the matrix $\langle fx^i,\, x^j\rangle_{L_2(0,1)}$ ($i = 1,2,\dots,n$; $j = 0,1,2,\dots,n-1$) is the Gramian of the linearly independent system $\{\mathcal{F}x^j\}_{j=0}^{n-1}$, where $\mathcal{F} = \sqrt{-xf}$ is nonzero, system (2.4.6) can be solved for the higher derivatives:

$$\frac{d^2u_{ni}}{dv^2} = \sum_{j=1}^{n}\Big[a_{ij}(v)\frac{du_{nj}}{dv} + b_{ij}(v)u_{nj}\Big] + \mathcal{F}_i(v), \qquad (2.4.8)$$

where $i = 1,2,\dots,n$, and a_{ij}, b_{ij}, $\mathcal{F}_i$ are given continuous functions.

Thus, we arrive at the problem (2.4.8), (2.4.7): Find u_{ni}, $i = 1,2,\dots,n$, that satisfy the system (2.4.8) and conditions (2.4.7). We shall prove that the problem (2.4.8), (2.4.7) has only trivial solution. Assume the contrary. Suppose that the homogeneous version of the problem (2.4.8), (2.4.7) has a nonzero solution $\bar{u}_n = (\bar{u}_{ni})$. Since (u_{ni}) satisfies the homogeneous version of system (2.4.8), it also satisfies (2.4.6) when $F_v = 0$. This can be easily verified by multiplying the homogeneous version of (2.4.8) by the matrix $\langle fx^i,\, x^j\rangle_{L_2(0,1)}$ ($i = 1,2,\dots,n$; $j = 0,1,\dots,n-1$).

Substitute $\bar{u}_n$ for $u_n = (u_{ni})$ in (2.4.6) with $F_v = 0$, multiply the jth equation of the system by $2(j+1)\bar{u}_{nj+1}$ and sum from 0 to $n-1$ with respect to j. We obtain

$$2\Big\langle \frac{\partial}{\partial v}(l\bar{u}_n),\, \bar{u}_{nx}\Big\rangle_{L_2(0,1)} = 0,$$

where

$$\bar{u}_n = \sum_{i=1}^{n} \bar{u}_{ni}x^i.$$

Consequently, since $\bar{u}_n \in \mathcal{M}$, $f(1,v) \le 0$, by virtue of (2.4.4) with $u_n = \bar{u}_n$, we have

$$\int_\Omega (\bar{u}_{nx}^2 + f_x \bar{u}_{nv}^2)\,\mathrm{d}\Omega \le 0,$$

i.e., $\bar{u}_n = 0$ because $f_x > 0$ and $\bar{u}_n = 0$ for $x = 0$, which contradicts the condition $\bar{u}_n \neq 0$. Therefore, for any continuous function F_v, system (2.4.6) has a unique solution which satisfies conditions (2.4.7) and is twice continuously differentiable. Note that since $f \in C^1(\overline{\Omega})$ and $f_x = 0$, there exists a number $\mu > 0$ such that $f_x > \mu > 0$.

We now estimate the solution u_n of (2.4.6), (2.4.7) in terms of F_v. For this purpose, we multiply the jth equation of (2.4.6) by $2(j+1)u_{nj+1}$ and sum from 0 to $n-4$ with respect to j. We have

$$2\left\langle \frac{\partial}{\partial v}(lu_n),\, u_{nx}\right\rangle_{L_2(0,1)} = 2\,\langle F_v,\, u_{nx}\rangle_{L_2(0,1)} \tag{2.4.9}$$

Then we estimate the right-hand side of (2.4.9) integrated over $(-1,1)$:

$$2\left|\int_\Omega F_v u_{nx}\,\mathrm{d}\Omega\right| \le 4\int_\Omega F_v^2\,\mathrm{d}\Omega + \frac{1}{4}\int_\Omega u_{nx}^2\,\mathrm{d}\Omega.$$

For $u_n = \bar{u}_n$, the integral of the left-hand side of (2.4.9) over $(-1,1)$ (as shown above) is greater than

$$\int_\Omega (u_{nx}^2 + f_x u_{nv}^2)\,\mathrm{d}\Omega.$$

Therefore, from (2.4.9) we obtain

$$\int_\Omega (u_{nx}^2 + f_x u_{nv}^2)\,\mathrm{d}\Omega \le 4\int_\Omega F_v^2\,\mathrm{d}\Omega + \frac{1}{4}\int_\Omega u_{nx}^2\,\mathrm{d}\Omega.$$

Since Ω is bounded, $f_x > \mu$, and $u_n = 0$ when $x = 0$, the last inequality yields

$$\|u_n\|_{H_1(\Omega)} \le C\,\|F_v\|_{L_2(\Omega)}, \tag{2.4.10}$$

where the constant $C > 0$ does not depend on n.

Using (2.4.10) it is easy to show that there exists a subsequence of $\{u_n\}$, which we again denote by $\{u_n\}$, such that $\{u_n\}$, $\{u_{nx}\}$, and $\{u_{nv}\}$ weakly converge in $L_2(\Omega)$ to $u(x,v)$, u_x, and u_v, respectively, as $n \to \infty$. Evidently, from the definition of the space $\overset{\circ}{H}_1(\Omega)$ it follows that $u = 0$ when $x = 0$ and $v = \pm 1$.

Multiplying the jth relation in (2.4.6) by an arbitrary function $\omega(v) \in C^2(-1,1)$, we transfer the derivative with respect to v onto the functions ωx^j. Integrating the obtained equality over $(-1,1)$, for $n-1 \geq j$ we have

$$\left\langle lu_n - F, \frac{\partial}{\partial v}(\omega x^j)\right\rangle = 0. \tag{2.4.11}$$

Passing to the limit in (2.4.11) as $n \to \infty$, we obtain

$$\left\langle lu - F, \frac{\partial}{\partial v}\eta\right\rangle = 0 \tag{2.4.12}$$

for any $\eta \in \overset{\circ}{H}_{1,2}(\Omega)$.

Putting $\lambda = lu - F$ we see from (2.4.12) that λ does not depend on x.

We have found a solution (u, λ) to Problem 2.4.1 such that $u \in \overset{\circ}{H}_1(\Omega)$ and $\lambda \in L_2(\Omega)$, which completes the proof of Theorem 2.4.2. □

In Anikonov and Bondarenko (1982), the authors establish the existence (in the small) and uniqueness of a solution in the class of analytic functions for the following inverse problem (under the condition $u_v(0,0,0) \neq 0$):

Determine the functions $u(x,v,t)$ and $f(x,v)$ from the equation

$$u_t + vu_x + fu_v = b(x,v,t),$$

provided that the solution $u(x,v,t)$ is known on the surface $x = 0$, i.e., $u(0,v,t) = u_0(v,t)$.

In what follows, we consider the question of solution uniqueness for a similar inverse problem for the stationary kinetic equation in the class of twice continuously differentiable functions.

Consider the equation

$$lu \equiv vu_x + fu_v + au = b(x,v)g(x) + C(x,v), \tag{2.4.13}$$

where f, a, b, C are given real-valued functions, and $g(x)$ is unknown.

Problem 2.4.2. Determine $u(x,v)$ and $g(x)$ from equation (2.4.13) and the condition

$$u(0,v) = u_0(v), \tag{2.4.14}$$

Theorem 2.4.3. *Suppose that $f, b \in C^1(\Omega)$, $f > 0$, $b > 0$, $b_v > 0$, $a > 0$, and a does not depend on v. Then Problem 2.4.2 has at most one solution $(u, g) \in C^2(\Omega) \times C(0, \infty)$.*

Proof. Let $(u, g) \in C^2(\Omega) \times C(x < 0)$ be a solution to Problem 2.4.2 with $C = u_0 = 0$. By (2.4.13)

$$\frac{1}{b}(vu_x + fu_v + au) = g(x).$$

Differentiating this equality with respect to v, we get

$$vu_{xv} + fu_{vv} + u_x + (f_v + a)u_v - \frac{b'_v}{b}(vu_x + fu_v + au) = 0 \qquad (2.4.15)$$

Let $\Omega_{\alpha,\beta}$ be a domain of the form

$$\Omega_{\alpha,\beta} = \{(x, v) : \frac{1}{\beta}(-\alpha + v^2) < x < 0\},$$

where the number $\alpha > 0$ is chosen so that $|vb'_v/b| < 1/2$ for $(x, v) \in \Omega_{\alpha,\beta}$, and $\beta > 1 + 4f$.

Multiply (2.4.15) by $2u_x \exp[\lambda(\beta x - v^2 + \alpha - \varepsilon)]$, where $0 < \varepsilon < \alpha$, $\lambda > 0$. It is easy to verify that

$$2u_x\Big(\frac{\partial}{\partial v}(lu) - \frac{b'_v}{b}(lu)\Big)\exp(\lambda\varkappa) = J(u) + \frac{\partial}{\partial x_1}U_1 + \frac{\partial}{\partial v}U_2, \qquad (2.4.16)$$

where

$$\varkappa = \beta x - v^2 + \alpha - \varepsilon,$$

$$J(u) = \exp(\lambda\varkappa)\Big[u_x^2 + 4\lambda f v u_x u_v + 2\lambda v^2 u_x^2 + f_x u_v^2$$
$$+ \lambda f \beta u_v^2 + 2au_x u_v - \frac{b'_v}{b}(vu_x + fu_v)u_x$$
$$+ \Big(\frac{b'_v}{b}a\Big)_x u^2 + \lambda\beta\frac{b'_v}{b}au^2\Big],$$

$$U_1 = \Big(fu_v^2 + \frac{b'_v}{b}au^2\Big)\exp(-\lambda\varkappa),$$

$$U_2 = (vu_x^2 + 2fu_v u_x)\exp(-\lambda\varkappa).$$

Since $\left|\frac{b'_v}{b}v\right| < \frac{1}{2}$ and $\beta > 1 + 4f$, we have

$$u_x^2 + 4\lambda f v u_x u_v + 2\lambda v^2 u_x^2 + f_x u_v^2$$

$$+ \lambda\beta f u_v^2 + 2a u_x u_v - v\frac{b'_v}{b}u_x^2 - \frac{b'_v}{b} f u_x u_v$$

$$\geq \frac{1}{2}u_x^2 - \lambda v^2 u_x^2 - 4\lambda f^2 u_v^2 + 2\lambda v^2 u_x^2 + f_x u_v^2$$

$$+ \lambda\beta f u_v^2 - \frac{1}{8}u_x^2 - 8a^2 u_v^2 - \frac{1}{8}u_x^2 - 8\Big(\frac{b'_v}{b}f\Big)^2 u_v^2$$

$$= u_x^2\Big(\frac{1}{4} + \lambda v^2\Big) + \Big[\lambda f(\beta - 4f) + f_x - 8a^2 - 8\Big(\frac{b'_v}{b}f\Big)^2\Big]u_v^2$$

$$\geq u_x^2\Big(\frac{1}{4} + \lambda v^2\Big) + \Big[\lambda f + f_x - 8a^2 - 8\Big(\frac{b'_v}{b}f\Big)^2\Big]u_v^2. \tag{2.4.17}$$

Let $\lambda > \lambda_0$, where $\lambda_0 = \max(\lambda_1, \lambda_2)$,

$$\lambda_1 = \max_{(x,v)\in\overline{\Omega}_{\alpha,\beta}} \Big[\frac{1}{f}\Big(1 + |f_x| + 8a^2 + 2\Big(\frac{b'_v}{b}f\Big)^2\Big)\Big],$$

$$\lambda_2 = \max_{(x,v)\in\overline{\Omega}_{\alpha,\beta}} \Big(1 + \Big|\Big(\frac{b'_v}{b}a\Big)_x\Big|\Big)\Big(\frac{b'_v}{b}a\Big)^{-1}.$$

From (2.4.16) and (2.4.17) we have

$$0 = 2u_x\Big[\frac{\partial}{\partial v}(lu) - \frac{b'_v}{b}(lu)\Big]\exp(\lambda\varkappa)$$

$$\geq \Big(\frac{1}{4}u_x^2 + u_v^2 + u^2\Big)\exp(\lambda\varkappa) + \frac{\partial}{\partial v}U_2 + \frac{\partial}{\partial x}U_1 \tag{2.4.18}$$

for $(x, v) \in \overline{\Omega}_{\alpha,\beta}$.

Integrating (2.4.18) over Ω and passing to the limit as $\lambda \to \infty$, we obtain

$$\int_{\Omega_{\alpha-\varepsilon,\beta}} u^2\, dx\, dv \leq 0$$

since $u(0, v) = 0$ and $\exp(\lambda\varkappa) = \exp(-\lambda\varepsilon)$ on $\partial\Omega_{\alpha,\beta} \setminus \{(x, v) : x = 0\}$. Thus, $u = 0$ in $\Omega_{\alpha-\varepsilon,\beta}$ and therefore $u = 0$ in $\Omega_{\alpha,\beta}$ since ε is arbitrary $(0 < \varepsilon < \alpha)$. Moreover, from (2.4.13) it follows that $g = 0$ in $[-\alpha, 0]$.

Having found $u(x,v)$ in $\Omega_{\alpha,\beta}$ and $g(x)$ in $[-\alpha,0]$, from (2.4.13) and (2.4.14) we can find $u(v,x)$ in $\Omega_{\alpha/2} = \{(x,v) : x \in [-\alpha/2,0],\ v \in \mathbb{R}^1\}$. Thus, we obtain the Cauchy problem for equation (2.4.13) in the domains

$$\Omega^+_{\alpha/2} = \left\{(x,v) : x \in (-\alpha/2,0),\ v > \sqrt{\frac{\beta+2}{2}\alpha}\right\}$$

and

$$\Omega^-_{\alpha/2} = \left\{(x,v) : x \in (-\alpha/2,0),\ -\infty < v < -\sqrt{\frac{\beta+2}{2}\alpha}\right\},$$

where equation (2.4.13) has no points of degeneracy.

After finding $u(x,v)$ in $\Omega_{\alpha/2}$, we can repeat the above arguments for the domain $\widetilde{\Omega}_\alpha = \{(x,v) : -\infty < x < -\alpha/2,\ v \in \mathbb{R}^1\}$, and so on. Hence $u = 0$ in Ω and $g = 0$ in $(-\infty, 0)$, which completes the proof. □

The next theorem follows from Theorem 2.4.3.

Theorem 2.4.4. *Suppose that the functions a, b, c, g in (2.4.13) are given and belong to C^1, $a > 0$, the function $f(x)$ is unknown and does not depend on v. Then the problem of determining the pair $(u(x,v), f(x))$ from equation (2.4.13) and condition (2.4.14) has at most one solution such that $(u,f) \in C^2(\Omega) \times C(-\infty,0)$, $u_v \neq 0$, $u_{vv} \neq 0$.*

Remark 2.4.2. Note that in Problem 2.4.2 we determine two functions, $u(x,v)$ and $g(x)$, from one function $u_0(v)$ (see also Theorem 2.4.4). This is possible due to the fact that equation (2.4.13) is degenerate for $v = 0$ (this fact is essential in the proof of Theorem 2.4.3).

Consider the equation

$$u_t + vu_x + f(x,v)u_v + a(x)u = b(x,v)g(x,t) + C, \tag{2.4.19}$$

where $(x,v) \in \Omega$, $t \in (0,\infty)$, and f, a, b, C are given real-valued functions; $g(x,t)$ is unknown and does not depend on v.

Problem 2.4.3. Determine the pair of functions $(u(x,v,t),\ g(x,t))$ from equation (2.4.19) and the data

$$u(0,v,t) = u_0(v,t) \tag{2.4.20}$$

$$u(x,v,0) = u_1(x,v) \tag{2.4.21}$$

Theorem 2.4.5. *Assume that the functions f, a, b do not depend on t and satisfy the assumptions of Theorem 2.4.3; moreover, $b > 0$, $b_v > 0$. Then Problem 2.4.3 has at most one solution $(u(x,v,t), g(x,t))$ such that $g \in C((-\infty,\infty) \times (0,\infty))$, $u(x,v,t)$ is twice continuously differentiable and $|u(x,v,t)| \leq K \exp(\alpha_0 t)$ for $t > t_0$, where the constants K, α_0, t_0 depend on $u(x,v,t)$.*

Proof. Assume that $u_0 = u_1 = C = 0$ and let (u, g) be a solution of Problem 2.4.3 that satisfies the conditions stated in Theorem 2.4.5. We apply the Laplace transform with respect to t to equation (2.4.19) and condition (2.4.20). As a result, we have

$$p\hat{u} + v\hat{u}_x + f\hat{u}_v + a\hat{u} = b\hat{g}(x,p), \tag{2.4.22}$$

$$\hat{u}(0,v,p) = 0, \tag{2.4.23}$$

where $\hat{u}$ and $\hat{g}$ are the images of u and g under the Laplace transform ($p = p_1 + ip_2$ is the Laplace transform parameter).

From the hypotheses of the theorem and the properties of the Laplace transform it follows that $\hat{u}(x,v,p)$ is analytic with respect to p in the strip $\mathrm{Re}\, p < -\alpha_0$. Therefore, if $\hat{u}(x,v,p) = 0$ for $p \in B$, where B is a unit ball lying in the strip $\mathrm{Re}\, p < -\alpha_0$, then $\hat{u}(x,v,p) = 0$ for all p such that $\mathrm{Re}\, p < -\alpha_0$. In what follows, we shall assume that $p \in B$.

Let $\hat{u} = \hat{u}_1 + i\hat{u}_2$, $\hat{g} = \hat{g}_1 + i\hat{g}_2$. Then (2.4.22) splits into two equations:

$$l\hat{u}_i + l_i\hat{u} = b\hat{g}_i, \qquad i = 1, 2,$$

where $l_1\hat{u} = p_1\hat{u}_1 - p_2\hat{u}_2$, $l_2\hat{u} = p_1\hat{u}_2 + p_1\hat{u}_1$, $\hat{u}_i(0,v,p) = 0$.

The following estimate similar to (2.4.18) can be obtained for $\hat{u}_i$ in $\Omega_{\alpha,\beta}$:

$$\begin{aligned} 0 &= 2\frac{\partial}{\partial x}\hat{u}_i\Big[\frac{\partial}{\partial v}(l\hat{u}_i) + \frac{\partial}{\partial v}(l_i\hat{u}) - \frac{b'_v}{b}(l\hat{u}_i + l_i\hat{u})\Big]\exp(\lambda\varkappa) \\ &\geq \Big[\hat{u}_{ix}^2\Big(\frac{1}{4} + \lambda v^2\Big) + \Big(\lambda f + f_x - 8a^2 - 2\Big(\frac{b'_v}{b}f\Big)^2\Big)\hat{u}_{iv}^2 + \lambda\beta\frac{b'_v}{b}a\hat{u}_i^2 \\ &\quad + \Big(\frac{b'_v}{b}a\Big)_x\hat{u}_i^2 + 2u_{ix}\frac{\partial}{\partial v}(l_i\hat{u}) - 2\hat{u}_{ix}\frac{b'_v}{b}l_i\hat{u}\Big]\exp(\lambda\varkappa) \\ &\quad + \frac{\partial}{\partial x}\widehat{U}_1^{(i)} + \frac{\partial}{\partial v}\widehat{U}_2^{(i)}, \end{aligned} \tag{2.4.24}$$

where

$$\widehat{U}_1^{(i)} = -\left(f\hat{u}_{iv}^2 + \frac{b'_v}{b}a\hat{u}_i^2\right)\exp(\lambda\varkappa),$$

$$\widehat{U}_2^{(i)} = (v\hat{u}_{ix}^2 + 2f\hat{u}_{iv}\hat{u}_{ix})\exp(\lambda\varkappa).$$

Let us now estimate the expressions

$$\left[2\hat{u}_{ix}\frac{\partial}{\partial v}(l_i\hat{u}) - 2\hat{u}_{ix}\frac{b'_v}{b}l_i\hat{u}\right]\exp(\lambda\varkappa).$$

We have

$$\begin{aligned}2\hat{u}_{1x}\frac{\partial}{\partial v}(l_1\hat{u})\exp(\lambda\varkappa) &= 2\hat{u}_{1x}(p_1\hat{u}_{1v} - p_2\hat{u}_{2v})\exp(\lambda\varkappa)\\ &\le \left(\frac{1}{32}\hat{u}_{1x}^2 + 32p_1^2\hat{u}_{1v}^2 + \frac{1}{32}\hat{u}_{1x}^2 + 32p_2^2\hat{u}_{2v}^2\right)\exp(\lambda\varkappa)\\ &= \left[\frac{1}{16}\hat{u}_{1x}^2 + 32(p_1^2\hat{u}_{1v}^2 + p_2^2\hat{u}_{2v}^2)\right]\exp(\lambda\varkappa),\end{aligned}$$

$$\begin{aligned}2\hat{u}_{1x}l_1\hat{u}\frac{b'_v}{b}\exp(\lambda\varkappa) &= 2\hat{u}_{1x}(p_1\hat{u}_1 - p_2\hat{u}_2)\frac{b'_v}{b}\exp(\lambda\varkappa)\\ &\le \left[\frac{1}{16}\hat{u}_{1x}^2 + 32(p_1^2\hat{u}_1^2 + p_2^2\hat{u}_2^2)\left(\frac{b'_v}{b}\right)^2\right]\exp(\lambda\varkappa),\end{aligned}$$

Hence

$$\begin{aligned}&2\hat{u}_{1x}\left[\frac{\partial}{\partial v}(l_1\hat{u}) - \frac{b'_v}{b}(l_1\hat{u})\right]\exp(\lambda\varkappa)\\ &\quad\le \left[\frac{1}{8}\hat{u}_{1x}^2 + 32(p_1^2\hat{u}_{1v}^2 + p_2^2\hat{u}_{2v}^2) + 32(p_1^2\hat{u}_1^2 + p_2^2\hat{u}_2^2)\left(\frac{b'_v}{b}\right)^2\right]\exp(\lambda\varkappa).\end{aligned}$$

Similarly,

$$\begin{aligned}&2\hat{u}_{2x}\left[\frac{\partial}{\partial v}(l_2\hat{u}) - \frac{b'_v}{b}(l_2\hat{u})\right]\exp(\lambda\varkappa)\\ &\quad\le \left[\frac{1}{8}\hat{u}_{2x}^2 + 32(p_1^2\hat{u}_{2v}^2 + p_2^2\hat{u}_{1v}^2) + 32(p_1^2\hat{u}_2^2 + p_2^2\hat{u}_1^2)\left(\frac{b'_v}{b}\right)^2\right]\exp(\lambda\varkappa).\end{aligned}$$

Using the obtained estimates, from (2.4.24) we have

$$0 = 2\sum_{i=1}^{2} \frac{\partial \hat{u}_i}{\partial x_i}\Big[\frac{\partial}{\partial v}(l\hat{u}_i + l_i\hat{u}) - \frac{b'_v}{b}(l\hat{u}_i + l_i\hat{u})\Big]\exp(\lambda\varkappa)$$

$$\geq \sum_{i=1}^{2}\Big\{\Big[\hat{u}_{ix}^2\Big(\frac{1}{8} + \lambda v^2\Big) + \Big(\lambda f + f_x - 8a^2 - 2\Big(\frac{b'_v}{b}f\Big)^2\Big)\hat{u}_{iv}^2$$

$$+ \lambda\beta\frac{b'_v}{b}a\hat{u}_i^2 + \Big(\frac{b'_v}{b}a\Big)_x\hat{u}_i^2 - 32\hat{u}_{iv}^2(p_1^2 + p_2^2)$$

$$- 32\Big(\frac{b'_v}{b}\Big)^2(p_1^2 + p_2^2)\hat{u}_i^2\Big]\exp(\lambda\varkappa) + \frac{\partial}{\partial x}\widehat{U}_1^{(i)} + \frac{\partial}{\partial v}\widehat{U}_2^{(i)}\Big\}.$$

By virtue of the last estimate, the final part of the proof is similar to that of Theorem 2.4.3. □

2.5. EQUATIONS OF THE BOLTZMANN TYPE

First, we analyze inverse problems for equations of the Boltzmann type.

Consider the equation

$$\frac{\partial u}{\partial t} + \{u, H\} + I(u) = \lambda(x, v, t), \tag{2.5.1}$$

in the domain $\Omega_T = \Omega \times (0, T)$, where $\{\cdot\,,\cdot\}$ is the Poisson bracket, H is a given function of x, v, and t,

$$I(u) = \int_G \int_{|\bar{n}|=1} B(v, w, \bar{n})[u(v^*)u(w^*) - u(v)u(w)]\,d\bar{n}\,dw.$$

Here $u(v) = u(x, v, t)$, $v^* = v^*(v, w, \bar{n})$, and $w^* = w^*(v, w, \bar{n})$ are given functions, λ is unknown and satisfies the equation

$$\sum_{i=1}^{n} \frac{\partial^2 \lambda}{\partial x_i \partial v_i} = 0, \tag{2.5.2}$$

$u(x, v, t)$ is the density of distribution of the number of particles in the phase space Ω, x is a coordinate vector in n-dimensional space, v and t

denote velocity and time, respectively; v and w are the velocities of the particles before collision, v^* and w^* are the velocities of the particles after collision, $d\bar{n}$ is the surface element of the unit sphere S^{n-1}, and Ω is the domain introduced in Section 2.2.

In what follows, we assume $v^*, w^* \in C^1(\widetilde{\Omega})$, where $\widetilde{\Omega} = \overline{G} \times \overline{G} \times S^{n-1}$, $\overline{G}$ is the closure of G, and for any $(v, w, n) \in \Omega$

$$\frac{\partial w^*}{\partial w} \neq 0, \qquad \frac{\partial v^*}{\partial w} \neq 0.$$

These conditions hold in most cases (see Klimantovich, 1982).

For $H = \frac{1}{2}|v|^2 + f(x)$ and $B = |v - w|$, (2.5.1) is the Boltzmann equation (see Vladimirov and Volovich, 1985; Klimantovich, 1982).

Problem 2.5.1. Find the pair of functions (u, λ) defined in Ω_T that satisfies equation (2.5.1), assuming that the Hamiltonian $H \in C^2(\Omega_T)$ is given and the trace of the solution u of (2.5.1) on the boundary $\partial\Omega_T$ is

$$u\big|_{\partial\Omega_T} = u_0. \tag{2.5.3}$$

We now introduce a set of functions $\mathfrak{M}(M_1, M_2)$. A function $F \in C^2(\Omega_T)$ belongs to $\mathfrak{M}(M_1, M_2)$ if and only if $F = 0$ for $v \notin G$ and the following two conditions hold:

$$4\pi \max_{(x,t)\in D\times(0,T)} \int_G \int_G \bigg\{ \int_{|\bar{n}|=1} B^2 \Big(F^2(v^*) \Big|\frac{\partial w}{\partial w^*}\Big| + F^2(w^*) \Big|\frac{\partial w}{\partial v^*}\Big| \Big)\, d\bar{n}$$
$$+ \Big(\int_{|\bar{n}|=1} B\, F(v)\, d\bar{n} \Big)^2 \bigg\}\, dv\, dw$$
$$+ \max_{(x,v,t)\in\Omega_T} \Big(\iint_{G,\bar{n}} B\, F(w)\, dw\, d\bar{n} \Big)^2 \le M_1;$$

$$4\pi \sum_{i=1}^{n} \max_{(x,t)\in D\times(0,T)} \int_G \int_G \bigg\{ \int_{|\bar{n}|=1} B^2 \Big(F^2_{x_i}(v^*) \Big|\frac{\partial w}{\partial w^*}\Big| + F^2_{x_i}(w^*) \Big|\frac{\partial w}{\partial v^*}\Big| \Big)\, d\bar{n}$$
$$+ \Big(\int_{|\bar{n}|=1} B\, F_{x_i}(v)\, d\bar{n} \Big)^2 \bigg\}\, dv\, dw$$
$$+ \max_{(x,v,t)\in\Omega_T} \sum_{i=1}^{n} \Big(\iint_{G,\bar{n}} |B\, F_{x_i}(w)|\, dw\, d\bar{n} \Big)^2 \le M_2;$$

Here and in what follows, by $G, \bar{n}$ we denote the set

$$G \times \{\, |\bar{n}| = 1\} = \{(v, \bar{n}) \,:\, v \in G,\, \bar{n} \in S^{n-1}\}.$$

Since Ω is bounded, there exists $C_0 > 0$ such that for any $u \in \overset{\circ}{H}_1 (\Omega)$

$$\|u\|^2_{L_2(\Omega)} \le C_0\, \|u\|^2_{H_1(\Omega)}.$$

Theorem 2.5.1. *Let a function $H \in C^2(\Omega_T)$ satisfy condition (2.2.4′) with $\alpha_1 \ge M$, $\alpha_2 \ge 8$, where $M = M_1 + C_0 M_2$. Then Problem 2.5.1 has at most one solution (u, λ) such that $u \in \mathfrak{M}$, $\lambda \in C^2(\Omega_T)$.*

Proof. Suppose that (u_1, λ_1) and (u_2, λ_2) are solutions to Problem 2.5.1 that both satisfy condition (2.5.3) with the same function u_0 and, moreover, $u_1, u_2 \in \mathfrak{M}$, $\lambda_1, \lambda_2 \in C^2(\Omega_T)$. Introducing $u = u_1 - u_2$ and $\lambda = \lambda_1 - \lambda_2$, we easily obtain the following relation for (u, λ):

$$Lu \equiv \frac{\partial u}{\partial t} + \{u, H\} + \iint\limits_{G,\bar{n}} B(v, w, \bar{n})[u_1(v^*)u(w^*) + u_2(w^*)u(v^*)$$
$$- u_1(v)u(w) - u_2(w)u(v)]\, dw\, d\bar{n} = \lambda(x, v, t), \tag{2.5.4}$$

$$u\Big|_{\partial\Omega_T} = 0, \tag{2.5.5}$$

where λ satisfies equation (2.5.2).

Since (2.5.2) holds for λ, we have

$$\sum_{i=1}^{n} u_{v_i} \lambda_{x_i} = \sum_{i=1}^{n} \frac{\partial}{\partial v_i}(u, \lambda_{x_i}). \tag{2.5.6}$$

Moreover, for $lu \equiv u_t + \{u, H\}$ we have

$$2\sum_{i=1}^{n} u_{v_i} \frac{\partial}{\partial x_i}(lu)$$
$$= \sum_{i,j=1}^{n} \Big(\frac{\partial^2 H}{\partial v_i \partial v_j} u_{x_i} u_{x_j} - \frac{\partial^2 H}{\partial x_i \partial x_j} u_{v_i} u_{v_j} \Big)$$
$$+ \sum_{i=1}^{n} [(u_{v_i} u_{x_i})_t + (u_{v_i} lu)_{x_i} - (u_{x_i} lu)_{v_i} + (u_{v_i} u_{x_i})_t]$$
$$+ \sum_{i,j=1}^{n} \Big[\Big(\frac{\partial H}{\partial v_j} u_{v_i} u_{x_i} \Big)_{x_j} - \Big(\frac{\partial H}{\partial x_j} \frac{\partial u}{\partial v_i} \frac{\partial u}{\partial x_i} \Big)_{v_j} \Big]. \tag{2.5.7}$$

We now estimate the expression

$$2\sum_{i=1}^{n} u_{v_i}\frac{\partial}{\partial x_i}(I_1 u),$$

where

$$I_1 u = \iint_{G,\bar{n}} B(v,w,\bar{n})u_1(v^*)u(w^*)\,d\bar{n}\,dw.$$

Since $\dfrac{\partial w^*}{\partial w} \neq 0$ for all $(v,w,\bar{n}) \in \widetilde{\Omega}$ and $u = 0$ for $v \notin G$, we have

$$\begin{aligned}
&\left| 2u_{v_i} \iint_{G,\bar{n}} Bu_1(v^*)u_{x_i}(w^*)\,dw\,d\bar{n} \right| \\
&\quad \le u_{v_i}^2 + \left(\iint_{G,\bar{n}} Bu_1(v^*)u(w^*)\,dw\,d\bar{n} \right)^2 \\
&\quad \le u_{v_i}^2 + \left(\iint_{|\bar{n}|=1,\,G'_{\bar{n},v}} Bu_1(v^*)u_{x_i}(w^*)\frac{\partial w}{\partial w^*}\,dw^*\,d\bar{n} \right)^2 \\
&\quad \le u_{v_i}^2 + 4\pi \int_G u_{x_i}^2(w^*)\,dw^* \int_{|\bar{n}|=1}\int_{G'_{\bar{n},v}} \left(Bu_1(v^*)\frac{\partial w}{\partial w^*} \right)^2 dw^*\,d\bar{n} \\
&\quad \le u_{v_i}^2 + 4\pi \int_G u_{x_i}^2\,dw \iint_{G,\bar{n}} B^2u_1^2(v^*)\left|\frac{\partial w}{\partial w^*}\right| dw\,d\bar{n},
\end{aligned} \tag{2.5.8}$$

where the domain $G'_{\bar{n},v}$ $(G''_{\bar{n},v})$ is the image of G under the mapping $w^* : G \to G'_{\bar{n},v}$ $(v^* : G \to G''_{\bar{n},v})$ for fixed $\bar{n}$ and v: $G'_{\bar{n},v} = w^*(G)$ $(G''_{\bar{n},v} = v^*(G))$. Similar arguments lead to the estimate

$$\begin{aligned}
&2u_{v_i} \iint_{G,\bar{n}} Bu_{1x_i}(v^*)u(w^*)\,dw\,d\bar{n} \\
&\qquad \le u_{v_i}^2 + 4\pi \int_G u^2(w)\,dw \iint_{G,\bar{n}} B^2u_{1x_i}^2(v^*)\left|\frac{\partial w}{\partial w^*}\right| dw\,d\bar{n}.
\end{aligned} \tag{2.5.9}$$

The following estimates can be obtained in exactly the same way:

$$\Big| 2u_{v_i} \iint\limits_{G,\bar n} B u_2(w^*) u_{x_i}(v^*)\, d\bar n\, dw \Big|$$
$$\le u_{v_i}^2 + 4\pi \int_G u_{x_i}^2(w)\, dw \iint\limits_{G,\bar n} B^2 u_2^2(w^*) \Big| \frac{\partial w}{\partial v^*} \Big|\, dw\, d\bar n,$$

$$\Big| 2u_{v_i} \iint\limits_{G,\bar n} B u_{2x_i}(w^*) u(v^*)\, d\bar n\, dw \Big|$$
$$\le u_{v_i}^2 + 4\pi \int_G u^2(w)\, dw \iint\limits_{G,\bar n} B^2 u_{2x_i}^2 \Big| \frac{\partial w}{\partial v^*} \Big|\, d\bar n\, dw,$$

$$2u_{v_i} \iint\limits_{G,\bar n} B u_1(v) u_{x_i}(w)\, dw\, d\bar n$$
$$\le u_{v_i}^2 + \int_G u_{x_i}^2(w)\, dw \int_G \Big(\int_{|\bar n|=1} B u_1(v)\, d\bar n \Big)^2 dw,$$

$$\Big| 2u_{v_i} \iint\limits_{G,\bar n} B u_{1x_i}(v) u(w)\, d\bar n\, dw \Big|$$
$$\le u_{v_i}^2 + \int_G u^2(w)\, dw \int_G \Big(\int_{|\bar n|=1} B u_{1x_i}(v)\, d\bar n \Big)^2 dw,$$

$$\Big| 2u_{v_i} \iint\limits_{G,\bar n} B u_2(w) u_{x_i}(v)\, dw\, d\bar n \Big| \le u_{v_i}^2 + u_{x_i}^2(v) \Big(\iint\limits_{G,\bar n} B u_{2x_i}(w)\, d\bar n\, dw \Big)^2,$$

$$\Big| 2u_{v_i} u(v) \iint\limits_{G,\bar n} B u_{2x_i}(w)\, dw\, d\bar n \Big| \le u_{v_i}^2 + u^2(v) \Big(\iint\limits_{G,\bar n} B u_{2x_i}(w)\, d\bar n\, dw \Big)^2. \tag{2.5.10}$$

As follows from (2.5.5) – (2.5.10) and the fact that $u_1, u_2 \in \mathfrak{M}$, it is easy to see that

$$(\alpha_2 - M)\int_{\Omega_T} |\nabla_x u|^2 \, d\Omega_T \le 2\sum_{i=1}^{n} \int_{\Omega_T} \frac{\partial u}{\partial v_i} \frac{\partial (Lu)}{\partial x_i} \, d\Omega_T = 0.$$

Hence, $u \equiv 0$ and therefore $u_1 = u_2$ in Ω_T, and (2.5.4) implies $\lambda = 0$, i.e., $\lambda_1 = \lambda_2$. This completes the proof of Theorem 2.5.1. □

Problem 2.5.2. Determine the functions $u(x,v)$ and $\lambda(x,v)$ defined in Ω that satisfy the stationary equation (2.5.1) and the conditions

$$\nabla u\Big|_{\partial\Omega} = u_0, \quad \nabla_x \lambda\Big|_{\Gamma_2} = \lambda_0, \tag{2.5.11}$$

assuming that the Hamiltonian $H \in C^2(\overline{\Omega})$ and $B \in C^1(\widetilde{\Omega})$ are given.

Theorem 2.5.2. *Under the hypotheses of Theorem 2.5.1, Problem 2.5.2 has at most one solution* (u, λ) *such that* $u \in \mathfrak{M}$, $\lambda \in C^2(\Omega)$, $u(x^0, v^0) = 0$, *where* $(x^0, v^0) \in \Omega$.

The proof of Theorem 2.5.2 is similar to that of Theorem 2.5.1.

Consider the linearized equation of the Boltzmann type

$$lu + \iint_{G,\bar n} [u(v^*) + u(w) - u(v) - u(w)] \times f_0(w) B(v, w, \bar n) \, dw \, d\bar n = \lambda(x, v, t), \tag{2.5.12}$$

with the right-hand side λ satisfying condition (2.5.2). Here $f_0(w)$ is the Maxwell distribution.

Problem 2.5.3. Determine the pair of functions (u, λ) that satisfies equation (2.5.12), provided that the trace of the solution u of (2.5.12) on Ω_T is known: $u\Big|_{\partial\Omega_T} = u_0$.

Suppose that $u = 0$ for $v \notin G$. Then we may assume that $G'_{\bar n, v} \subset G$ for the integral in (2.5.12):

$$\iint_{G,\bar{n}} u(w^*) f_0(w) B(v, w, \bar{n}) \, dw \, d\bar{n}$$
$$= \iint_{G,\bar{n}} u(w^*) b'(w^*, \bar{n}, v) \frac{\partial w}{\partial w^*} \, dw^* \, d\bar{n},$$

where

$$b'(w^*, \bar{n}, v) = \begin{cases} f_0(w) B(v, w, \bar{n}), & \text{if } w^* \in G'_{\bar{n},v}, \\ 0, & \text{if } w^* \in G'_{\bar{n},v} \end{cases}$$

In exactly the same way we obtain

$$\iint_{G,\bar{n}} u(v^*) f_0(w) B(v, w, \bar{n}) \, dw \, d\bar{n}$$
$$= \iint_{G,\bar{n}} u(v^*) b''(v^*, \bar{n}, v) \frac{\partial w}{\partial v^*} \, dv^* \, d\bar{n},$$

where

$$b''(v^*, \bar{n}, v) = \begin{cases} f_0(w) B(v, w, \bar{n}), & \text{if } v^* \in G'_{\bar{n},v}, \\ 0, & \text{if } v^* \in G'_{\bar{n},v} \end{cases}$$

Then equation (2.5.12) can be written in the form

$$L_a u \equiv lu + \iint_{G,\bar{n}} a(v, w^*, \bar{n}) u(w^*) \, dw^* \, d\bar{n}$$
$$+ u(v) \iint_{G,\bar{n}} f_0(w) B(v, w, \bar{n}) \, dw \, d\bar{n} = \lambda(x, v, t),$$

where

$$a(v, w^*, \bar{n}) = \frac{\partial w}{\partial w^*} (b'(w^*, \bar{n}, v) + b''(w^*, \bar{n}, v)) + f_0(w) B(v, w^*, \bar{n}),$$

and $\dfrac{\partial w}{\partial w^*}$ is the Jacobian of the transformation $w \to w^*$. Set

$$\beta = \max_{v \in G} \left| \iint_{G,\bar{n}} f_0(w) B(v, w, \bar{n}) \, dw \, d\bar{n} \right| + 4\pi \iint_{G,\bar{n}} a^2(v, w^*, \bar{n}) \, dw^* \, d\bar{n},$$

and let $H_{2,1}(\Omega_T)$ be the set of all real-valued functions $u(x, v, t) \in L_2(\Omega)$ with generalized derivatives u_{x_i}, u_{v_i}, $u_{x_i v_j}$, $u_{x_i x_j}$, $i, j = 1, 2, \dots, n$, that also belong to $L_2(\Omega)$. (The space $H_{1,2}(\Omega)$ was introduced in Section 2.2).

Theorem 2.5.3. *Assume that $H \in C^2(\overline{\Omega}_T)$ satisfies condition (2.3.5) with $\alpha_1 > \beta$ and $\alpha_2 > 2$. Then there exists a solution (u, λ) of Problem 2.5.3 in $H_1(\Omega_T) \times L_2(\Omega_T)$, which is unique in $H_{2,1}(\Omega_T) \times L_2(\Omega_T)$.*

Proof. a) We establish the uniqueness of a solution to Problem 2.5.3 in the case where $(u, \lambda) \in C^2(\overline{\Omega}_T) \times C^2(\Omega_T)$. If $(u, \lambda) \in H_{2,1}(\Omega_T) \times L^2(\Omega_T)$, then the proof of uniqueness is similar to that of Theorem 2.4.1.

By (2.5.2) and (2.5.6) we have

$$\sum_{i=1}^{n} 2u_{v_i} \frac{\partial}{\partial x_i}(L_a u) = 2 \sum_{i=1}^{n} \frac{\partial}{\partial v_i}(u \lambda_{x_i}). \tag{2.5.13}$$

The following estimates hold:

$$\Big| 2u_{v_i} \iint\limits_{G,\bar{n}} u_{x_i}(w^*) a(v, w^*, \bar{n})\, dw\, d\bar{n} \Big|$$
$$\leq u_{v_i}^2 + 4\pi \int_G u_{x_i}^2(w)\, dw \iint\limits_{G,\bar{n}} u_{x_i}(w^*) a^2(v, w^*, \bar{n})\, dw\, d\bar{n};$$

$$\Big| 2u_{v_i} u_{x_i} \iint\limits_{G,\bar{n}} f_0(w) B(v, w, \bar{n})\, dw\, d\bar{n} \Big|$$
$$\leq u_{v_i}^2 + \Big(\iint\limits_{G,\bar{n}} f_0(w) B(v, w, \bar{n})\, d\bar{n}\, dw \Big)^2 u_{x_i}^2. \tag{2.5.14}$$

Using these estimates and equalities (2.5.7), (2.5.13), it is easy to obtain

$$(\alpha_1 - \beta) \int_{\Omega_T} |\nabla_x u|^2\, d\Omega_T \leq 2 \int_{\Omega_T} \sum_{i=1}^{n} u_{v_i} \frac{\partial}{\partial x_i}(L_a u)\, d\Omega_T = 0, \tag{2.5.15}$$

i.e., $u = 0$ in Ω_T, and (2.5.12) implies $\lambda = 0$ on Ω_T. Thus, the uniqueness part of the proof is complete.

b) We now turn to the existence part of the theorem. To this end, we first prove the same assertion for the following problem (provided that $\partial \Omega_T$

is sufficiently smooth and the concordance conditions for the data of the problem are satisfied).

Problem 2.5.3′. Determine the pair (u, λ) from the equation

$$L_a u - F = \lambda(x, v, t),$$

provided that $F \in H^2(\Omega_T)$ and the trace of the solution u on the boundary $\partial\Omega_T$ is given:

$$u\big|_{\partial\Omega_T} = 0.$$

Consider the Dirichlet problem

$$\sum_{i=1}^{n} \frac{\partial^2}{\partial x_i \partial v_i}(L_a u - F) = 0, \tag{2.5.16}$$

$$u\big|_{\partial\Omega_T} = 0. \tag{2.5.17}$$

To prove the existence theorem for the problem (2.5.16), (2.5.17), we use the Galerkin method. Let $\{\omega_i(x, v, t)\}_{i=1}^{\infty}$ be a basis in the space $\overset{\circ}{H}_{2,1}(\Omega_T) = H_{2,1}(\Omega_T) \cap \overset{\circ}{H}_1(\Omega_T)$ such that $\omega_i \in C^3(\Omega_T)$. Then $u_N = (\alpha_{N1}, \dots, \alpha_{NN})$ is determined from the linear system

$$\Big\langle \sum_{i=1}^{n} \frac{\partial^2}{\partial v_i \partial x_i}(L_a u_N), \, \omega_j \Big\rangle = \langle \mathcal{F}, \omega_j \rangle, \quad 1 \le j \le N, \tag{2.5.18}$$

where

$$u_N = \sum_{j=1}^{n} \alpha_{Nj}\omega_j, \quad \mathcal{F} = \sum_{i=1}^{n} \frac{\partial^2 F}{\partial x_i \partial v_i}.$$

For any $\mathcal{F} \in L_2(\Omega_T)$ the system (2.5.18) has a unique solution u_N since the homogeneous version of (2.5.18) has only trivial solution. This can be proved by contradiction with the use of estimate (2.5.15) (in exactly the same way as in the proof of Theorem 2.2.2). To establish an *a priori* estimate for u_N, we multiply the jth equation of (2.5.18) by $-2\alpha_{Nj}$ and sum from 1 to N with respect to j. Then, using (2.5.7), (2.5.14), and the condition that

$u_N = 0$ on $\partial\Omega_T$, from (2.5.18) we obtain

$$-2\langle \mathcal{F}, u_N\rangle = -2\sum_{i=1}^{n}\Big\langle \frac{\partial^2}{\partial x_i \partial v_i}(L_a u_N),\, u_N\Big\rangle$$

$$= 2\sum_{i=1}^{n}\Big\langle \frac{\partial}{\partial x_i}(L_a u_N),\, u_{Nv_i}\Big\rangle$$

$$\geq \int_{\Omega_T} \big((\alpha_1 - \beta)|\nabla_x u_N|^2 + (\alpha_2 - 2)|\nabla_v u_N|^2\big)\, \mathrm{d}\Omega_T,$$

which gives

$$\int_{\Omega_T} |\nabla_{x,v} u_N|^2\, \mathrm{d}\Omega_T \leq C,$$

where C is independent of N.

Using the obtained *a priori* estimate for u_N, one can prove the existence of a solution to Problem 2.5.3 in exactly the same way as in the proof of Theorem 2.2.2. □

Problem 2.5.4. Find a pair of functions (u, λ) satisfying equation (2.5.12) and conditions (2.5.11).

Theorem 2.5.4. *Under the hypotheses of Theorem 2.5.3, Problem 2.5.4 has at most one solution* $(u, \lambda) \in H_{2,1}(\Omega_T) \times L_2(\Omega_T)$.

The proof is similar to the proof of uniqueness in Theorem 2.5.3.

Consider the linear Boltzmann equation in Ω_T (see Cercignani, 1975):

$$lu + \iint_G K(v, v')u(v')\, \mathrm{d}v' + \nu(v)u(v) = \lambda(v), \tag{2.5.19}$$

where $K(v, v')$, $\nu(v)$ are known continuous functions and λ satisfies (2.5.2).

Problem 2.5.5. Determine the pair of functions (u, λ) from equation (2.5.19) and the condition $u\big|_{\partial\Omega_T} = u_0$.

Let

$$\beta = \max_{v\in G} |\nu(v)| + 4\pi \iint_{G,\bar{n}} K^2\, \mathrm{d}x\, \mathrm{d}v\, \mathrm{d}\bar{n}.$$

Theorem 2.5.5. *Under the hypotheses of Theorem 2.5.3, there exists a solution* (u, λ) *of Problem 2.5.5 in* $H_1(\Omega_T) \times L_2(\Omega_T)$*, which is unique in* $H_{2,1}(\Omega_T) \times L_2(\Omega_T)$.

Problem 2.5.6. Determine the pair of functions (u, λ) from (2.5.19) and conditions (2.5.11).

Theorem 2.5.6. *Under the hypotheses of Theorem 2.5.3, Problem 2.5.6 has at most one solution* $(u, \lambda) \in H_{2,1}(\Omega_T) \times L_2(\Omega_T)$.

Theorems 2.5.5 and 2.5.6 are proved analogously to Theorem 2.5.3.

2.6. THE VLASOV SYSTEM

1. Formulation of the problems. Suppose that a domain D of the n-dimensional Euclidean space is filled with plasma of rather low density. Then, in the model of plasma in the approximation of self-coordinated field proposed by A. A. Vlasov, the distribution functions for ions and electrons satisfy the following system (the Vlasov system; see Samarskii, 1980):

$$l_{\pm}^{\varphi} u^{\pm} \equiv \frac{\partial u^{\pm}}{\partial t} + \frac{1}{m_{\pm}} \langle p, \nabla_x u^{\pm} \rangle \mp \langle \nabla_x \varphi, \nabla_p u^{\pm} \rangle = g^{\pm}(x, p, t),$$

$$-\Delta\varphi = 4\pi\rho, \quad \rho = \int_{\mathbb{R}^n} (u^+ - u^-)\, dp, \tag{2.6.1}$$

where $x \in D$, $p \in \mathbb{R}^n$, $t > 0$, u^+ (u^-) is the function of distribution of ions (electrons) in the phase space, m_+ (m_-) is the ion (electron) mass, e is the charge of electron, and φ is the potential of self-coordinated electric field. We are interested in the following inverse problems.

Problem 2.6.1. Find the functions $u^{\pm}$, g^-, φ defined in the domain $\Omega = \{(x, \rho, t) : x \in D, p \in \mathbb{R}^n, t \in (0, T)\}$ that satisfy (2.6.1) and the condition

$$u^{\pm}\big|_{\partial\Omega} = u_0^{\pm}, \qquad \varphi\big|_{\partial D \times (0,T)} = \varphi_0, \tag{2.6.2}$$

g^+ is known, and g^- satisfies the equation

$$\sum_{i=1}^{n} \frac{\partial^2 g^-}{\partial x_i \partial p_i} = 0, \tag{2.6.3}$$

Problem 2.6.2. Find the functions $u^\pm$, φ, g^- defined in Ω that satisfy (2.6.1) and the conditions:

$$u^\pm\big|_{t=0} = u_0^\pm, \quad u^-\big|_{t=T} = u_T^-, \quad \varphi\big|_{\partial D\times(0,T)} = \varphi_0, \tag{2.6.4}$$

provided that g^- satisfies (2.6.3), g^+ is known, and for $x \in \partial D$

$$\begin{aligned} u^\pm(x,p,t) &= u^\pm(x, p - 2\bar{n}\langle\bar{n}, p\rangle, t), \\ g^-(x,p,t) &= g^-(x, p - 2\bar{n}\langle\bar{n}, p\rangle, t), \end{aligned} \tag{2.6.5}$$

where $\bar{n}$ is the unit outward normal to the surface of a solid body, directed inside the domain, and $\langle\cdot,\cdot\rangle$ is the scalar product in $\mathbb{R}^n$.

Problem 2.6.3. Find $u^\pm$, φ, $g^\pm$ satisfying (2.6.1) such that

$$u^\pm(0,'x,p,t) = u_0^\pm('x,p,t), \tag{2.6.6}$$

$$\varphi(0,'x,t) = \varphi_0, \quad \varphi'_{x_1}(0,'x,t) = \varphi_1, \tag{2.6.7}$$

$$u^\pm(x,p,0) = u_1^\pm(x,p). \tag{2.6.8}$$

Condition (2.6.5) is physically interpreted as the condition of regular reflection of particles from the solid body surface. The momentum component that is perpendicular to the surface only changes sign after the collision of a particle and the solid body, while the tangent components remain unchanged (see Silin, 1971).

2. We introduce the following notation. Let $\gamma_1 > 0$ be such that

$$\forall \varphi \in \overset{\circ}{H}_1(D) \quad \int_D \varphi_{x_1}^2\, dx \geq \gamma_1 \int_D \varphi^2\, dx,$$

and choose $\gamma_2 > 0$ with the property $8\pi/\gamma_2 < \gamma_1$, and γ_3 such that

$$\forall \varphi \in \overset{\circ}{H}_1(D) \quad \int_G \varphi_{p_1}^2\, dG > \gamma_3 \int_G \varphi^2\, dG,$$

where $G \subset \mathbb{R}^n$ is a bounded Lebesgue measurable set.

Let $\alpha > 4\pi\gamma_2/\gamma_3$, $M < m_-ne/32$, and let $\mathfrak{M}_{\alpha,M}$ be the set of all pairs $(u^\pm, \varphi)$ with the following properties:

$$u^\pm \in C^2(\Omega) \cap H^2(\Omega), \quad \varphi \in C^2(D);$$

$$|\nabla_p u^\pm| \leq M, \quad |\nabla_p u_{p_1}^-| < M, \quad u^\pm = 0 \quad \text{for } p \notin G;$$

$$\forall \xi \in \mathbb{R}^n \quad \sum_{i,j=1}^n \varphi_{x_i x_j} \xi^i \xi^j \geq \alpha |\xi|^2.$$

We take γ_4 such that $1 - (T-t)M^2 e\gamma_4 > 0$ and β such that

$$\beta(1 - \gamma_4(T-t)eM^2) - 4\pi\gamma_2 e \operatorname{mes} G > 0;$$

furthermore, let $(T-t)$ be sufficiently small to provide $\beta(T-t)/\gamma_4 < 3/8$.

Theorem 2.6.1. *Under the specified conditions, Problem 2.6.1 has at most one solution $(u^\pm, \varphi, g^-)$ such that $(u^\pm, \varphi) \in \mathfrak{M}_{\alpha,M}$ and $g^- \in C^2(\Omega)$.*

Proof. Suppose that there are two vector functions, $(u_1^\pm, \varphi_1, g_1^-)$ and $(u_2^\pm, \varphi_2, g_2^-)$, satisfying (2.6.1) – (2.6.3) and such that $(u_i^\pm, \varphi_i) \in \mathfrak{M}_{\alpha,M}$ and $g_i^- \in C^2(\Omega)$, $i = 1, 2$. Let $u^\pm = u_2^\pm - u_1^\pm$, $\varphi = \varphi_2 - \varphi_1$, and $g^- = g_2^- - g_1^-$. Then for $(u^\pm, \varphi, g^-)$ we have

$$l_\pm^{\varphi_1} u^\pm \mp e\langle \nabla_x \varphi, \nabla_p u_2^\pm \rangle = g^\pm, \tag{2.6.9}$$

$$-\Delta\varphi = 4\pi\rho, \quad \rho = \int_{\mathbb{R}^n} (u^+ - u^-)\,dp, \quad g^+ = 0, \tag{2.6.10}$$

and $(u^\pm, \varphi)$ satisfies the homogeneous version of condition (2.6.2).

We now apply the operator $2\sum_{i=1}^n u_{x_i}^- \dfrac{\partial}{\partial p_i}$ to equation (2.6.9) written for the function u^-. We obtain (see Anikonov and Amirov, 1983)

$$\begin{aligned} \frac{1}{m_-}|\nabla_x u^-|^2 &+ e\sum_{i,j=1}^n \varphi_{1x_ix_j} u_{p_i}^- u_{p_j}^- \\ &+ 2e\sum_{i=1}^n u_{x_i}^- \sum_{i,j=1}^n \varphi_{x_i} u_{2p_ip_j}^- + \operatorname{div} u^- = 0, \end{aligned} \tag{2.6.11}$$

$$\begin{aligned} \operatorname{div} u^- = \sum_{i,j=1}^n \big[&(p_i u_{x_i} u_{x_j} + f_i u_{p_i} u_{x_j})_{p_j} + (f_i u_{p_i} u_{x_j})_{p_i} \\ &+ (p_i u_{p_j} u_{x_j})_{x_i} + (p_i u_{x_i} u_{p_j} + f_i u_{p_i} u_{p_j})_{x_j} \big] \\ + \sum_{j=1}^n &\frac{\partial}{\partial t}(u_{p_j} u_{x_i}). \end{aligned}$$

Multiply equation (2.6.9) written for the function u^+ by $2\beta(T-t)u^+$:

$$\beta(u^+)^2 - 2\beta(T-t)e\sum_{i=1}^n \varphi_{x_i} u_{2p_i}^+ u^+ + \operatorname{div} u^+ = 0,$$

$$\operatorname{div} u^+ = \big[\beta(T-t)(u^+)^2\big]_t + \sum_{i=1}^{n}\Big\{\frac{1}{m_+}(p_i(u^+)^2)_{x_i} + e(\varphi_{x_i}(u^+)^2)_{p_i}\Big\}. \tag{2.6.12}$$

Multiply equation (2.6.10) by φ. We have

$$|\nabla_x\varphi|^2 + \operatorname{div}\varphi = 4\pi\varphi\int_{\mathbb{R}^n}(u^+ - u^-)\,dp \tag{2.6.13}$$

where $\operatorname{div}\varphi = \sum\limits_{i=1}^{n}(\varphi_{x_i}\varphi)_{x_i}$.

Consider the expressions

$$4\pi\varphi\int_{\mathbb{R}^n}(u^+ - u^-)\,dp \le \frac{2\pi}{\gamma_2}\varphi^2 + 2\gamma_2\pi \operatorname{mes} G\int_{\mathbb{R}^n}\big((u^+)^2 + (u^-)^2\big)\,dp;$$

$$\begin{aligned} 2e\sum_{i=1}^{n}u^-_{x_i}\Big(\sum_{j=1}^{n}\varphi_{x_j}u^-_{2p_ip_j}\Big) &= 2e\sum_{j=1}^{n}\varphi_{x_j}\Big(\sum_{i=1}^{n}u^-_{2p_ip_j}u^-_{x_i}\Big) \\ &\le \frac{e}{8}|\nabla_x\varphi|^2 + 16e\sum_{j=1}^{n}\Big(\sum_{i=1}^{n}u^-_{2p_ip_j}u^-_{x_i}\Big)^2 \\ &\le \frac{e}{8}|\nabla_x\varphi|^2 + 32e\,|\nabla_x u^-|^2\sum_{i,j=1}^{n}(u^-_{p_ip_j})^2 \end{aligned} \tag{2.6.14}$$

Summing equalities (2.6.11) – (2.6.13), we get

$$\begin{aligned} &\frac{1}{2}|\nabla_x\varphi|^2 + \big[\beta(1-\gamma_4(T-t)eM^2) - 4\pi \operatorname{mes} G\gamma_2 e\big](u^+)^2 \\ &\quad + \Big(\frac{1}{m_-} - 32eM^2n\Big)|\nabla_x u^-|^2 + e\Big(\alpha - \frac{4\pi\gamma_2}{\gamma_3}\Big)(u^-)^2 \\ &\quad + \operatorname{div} u^+ + \operatorname{div} u^- + \operatorname{div}\varphi \le 0 \end{aligned} \tag{2.6.15}$$

With the use of (2.6.2) and (2.6.14), integration of inequality (2.6.15) over Ω yields $\varphi = u^\pm = 0$, and (2.6.9) implies $g^- = 0$. This completes the proof of Theorem 2.6.1. □

A similar theorem holds for Problem 2.6.2, with the domain D of the form $D = \{x \in \mathbb{R}^n : -a_i < x_i < a_i,\, a_i > 0\}$. The proof is almost the same as that of Theorem 2.6.1. The only difference is that it is necessary to use the condition (2.6.5), which means that the functions $u^{\pm}$ and $g^{\pm}$ are even with respect to p_i on the hyperplanes $x_1 = \pm a_i$. The terms in (2.6.11) that are divergent with respect to x vanish after the integration.

3. In this part of the section we assume that the domain D has the following property: for any $x \in D$ there is a cone $K_x \subset D$ containing x, with base lying in the plane $x_1 = 0$.

Let

$$\Omega = \{(x, p, t) : x \in D,\, p \in \mathbb{R}^n,\, t \in \mathbb{R}^1\};$$

$$\mathfrak{M} = \{u : u \in C^2(\Omega),\, u = 0 \text{ if } p \notin G\};$$

$$\Omega(\gamma, \delta) = \{(x, p, t) : p \in \mathbb{R}^n,\, x_1 > 0,\; 0 < \delta x_1 + \frac{1}{2}\sum_{i=2}^{n}(x_i - x_i^0)^2 + \frac{t^2}{2} < \gamma\}.$$

Assume that $\delta_0 > 0$ and $\delta_1 > 0$ are some numbers, R is the radius of the smallest ball centred at the origin of $\mathbb{R}^n$ and containing the domain G, and δ satisfies the condition

$$\delta > (\delta_0 + (n-1)M + R^2 + 1)/\delta_1,$$

where $M > 0$, and $\Phi = \{\varphi : \varphi \in C^2,\, \varphi_{x_1} < -\delta_1,\, |\varphi_{x_i}| \le M\}$.

Lemma 2.6.1. *Assume that $x^0 = (0, x_2^0, \dots, x_n^0) \in D$, and let*

$$\bar{l}_{\pm}^{\varphi} u \equiv u_t + \frac{1}{m_{\pm}}\langle p, \nabla_x u\rangle + e\langle \nabla_x \varphi, \nabla_p u\rangle,$$

$$\psi(x, t) = \delta x_1 + \frac{1}{2}\sum_{i=2}^{n}(x_i - x_i^0)^2 + \frac{1}{2}t^2 + \alpha,$$

where $\alpha > 0$, $\alpha + \gamma < 1$. If $-\varphi_{x_1} \ge \delta_1 > 0$, $\lambda > 0$, $\nu > 0$, then every function $u \in C^2(\overline{\Omega}(\gamma, \delta))$ satisfies the following estimate in $\Omega(\gamma, \delta)$:

$$\psi^{\nu+1}(\bar{l}_{\pm}^{\varphi} u)^2 \exp(2\lambda\psi^{-\nu}) \ge D_{\pm}(u) + \delta\lambda\nu u^2 \exp(2\lambda\psi^{-\nu}), \qquad (2.6.16)$$

where

$$D_{\pm}(u) = \lambda\nu \frac{\partial}{\partial t}\Big\{\Big[\delta p_1 + \sum_{i=2}^{n}(p_i(x_i - x_i^0) + m_{\pm}t)\Big]u^2 \exp(2\lambda\psi^{-\nu})\frac{1}{m_{\pm}}\Big\}$$

$$+ \lambda\nu \sum_{i=1}^{n} \frac{\partial}{\partial p_i}\Big\{\Big[p_1 \delta\varphi_{x_1} + \sum_{j=2}^{n} p_j \varphi_{x_j}(x_j - x_j^0)$$

$$+ t\varphi_{x_i} m_{\pm}\Big]\frac{1}{m_{\pm}}u^2 \exp(2\lambda\psi^{-\nu})\Big\}$$

$$+ \lambda\nu \sum_{i=1}^{n} \frac{\partial}{\partial x_i}\Big\{\Big[\delta p_1 p_i + p_i \sum_{j=2}^{n}(p_j \psi_{x_j} + p_i m_{\pm} t)\frac{1}{m_{\pm}^2}u^2 \exp(2\lambda\psi^{-\nu})\Big]\Big\}.$$

Proof. Introducing the function $v = \exp(\lambda\psi^{-\nu})u$, we have

$$u_{x_i} = (v_{x_i} + \lambda\nu\psi^{-\nu-1}\psi_{x_i} v)\exp(-\lambda\psi^{-\nu}),$$

$$u_{p_i} = v_{p_i}\exp(-\lambda\psi^{-\nu}), \quad u_t = (v_t + \lambda\nu\psi^{-\nu-1}\psi_t v)\exp(-\lambda\psi^{-\nu}).$$

Therefore,

$$(\bar{l}_{\pm}^{\varphi} u)^2 \exp(2\lambda\psi^{-\nu})\psi^{\nu+1}$$

$$= \Big[\bar{l}_{\pm}^{\varphi} v + \lambda\nu\Big(\frac{\delta p_1}{m_{\pm}} + \sum_{i=2}^{n}\frac{p_i\psi_{x_i}}{m_{\pm}} + \psi_t\Big)\psi^{-\nu-1}v\Big]^2 \psi^{\nu+1}$$

$$\geq 2\lambda\nu v \bar{l}_{\pm}^{\varphi} v\Big(\frac{\delta p_1}{m_{\pm}} + \sum_{i=2}^{n}\frac{p_i\psi_{x_i}}{m_{\pm}} + \psi_t\Big)$$

$$= -\lambda\nu v^2\Big[\frac{\delta\varphi_{x_1}}{m_{\pm}} + \sum_{i=2}^{n}\frac{p_i^2}{m_{\pm}} + \sum_{i=1}^{n}(x_i - x_i^0)\varphi_{x_i} + 1\Big]$$

$$+ \lambda\nu\frac{\partial}{\partial t}\Big[\Big(\frac{\delta p_1}{m_{\pm}} + \sum_{i=2}^{n}\frac{p_i\psi_{x_i}}{m_{\pm}} + t\Big)v^2\Big]$$

$$+ \lambda\nu\sum_{i=1}^{n}\frac{\partial}{\partial x_i}\Big(\frac{\delta p_1 p_i}{m_{\pm}^2} + \frac{p_i}{m_{\pm}^2}\sum_{j=2}^{n} p_j\psi_{x_j} + \frac{p_i}{m_{\pm}}t\Big)v^2$$

$$+ \lambda\nu\sum_{i=1}^{n}\frac{\partial}{\partial p_i}\Big[\frac{p_1}{m_{\pm}}\delta\varphi_{x_i} + \sum_{j=2}^{n}\frac{p_j}{m_{\pm}}\psi_{x_j}\varphi_{x_i} + t\varphi_{x_i}v^2\Big]$$

Returning to the function $u(x,p,t)$ in the right-hand side of the last equality, it is easy to obtain the estimate (2.6.16) $\square$

Lemma 2.6.2. *Let $\nu > (|\nabla_x\psi|^2(n-1)+2|\nabla_{'x}\psi|^2+1)\delta^{-4}$ and $\lambda \geq \lambda_0 > 0$ in $\Omega(\gamma,\delta)$, where $'x = (x_2,\dots,x_n)$ and λ_0 is a sufficiently large number. Then*

$$\begin{aligned}\psi^{\nu+1}&(\Delta\varphi)^2\exp(2\lambda\psi^{-\nu})\\ &\geq -2\lambda\nu\sum_{i=2}^{n}\varphi_{x_i}^2\exp(2\lambda\psi^{-\nu})\\ &\quad+\delta^4\lambda^3\nu^4\psi^{-2(\nu+1)}\varphi^2\exp(2\lambda\psi^{-\nu})+A(\varphi),\end{aligned}$$

where

$$\begin{aligned}A(\varphi)=\sum_{i,j=1}^{n}\Big\{&4\lambda\nu\big[\psi_{x_i}(\varphi_{x_i}-\lambda\nu\psi^{-\nu-1}\psi_{x_i}\varphi)\\ &\times(\varphi_{x_j}-\lambda\nu\psi^{-\nu-1}\psi_{x_j}\varphi)\exp(2\lambda\psi^{-\nu})\big]_{x_j}\\ &-2\lambda\nu\big[\psi_{x_i}(\varphi_{x_j}-\lambda\nu\psi^{-\nu-1}\psi_{x_j}\varphi)^2\exp(2\lambda\psi^{-\nu})\big]_{x_i}\Big\}\\ +\sum_{i=1}^{n}&2\lambda^2\nu^2\big[\psi_{x_i}\varphi^2\exp(2\lambda\psi^{-\nu})(\psi^{-2\nu-2}|\nabla_x\psi|^2\lambda\nu\\ &-(\nu+1)\psi^{-\nu-2}|\nabla_x\psi|^2+\psi^{-\nu-1}\psi_{x_jx_j})\big]_{x_i}\end{aligned}$$

Lemma 2.6.2 can be proved by the method used in the proof of Lemma 2 (Lavrent'ev, Romanov, and Shishatskii (1980), p. 82) with certain modifications. See also Lemma 4.2.1 of the present book.

Lemma 2.6.3. *For any $\varphi \in C^2(\overline{\Omega}(\gamma,\delta))$, the following equality holds:*

$$\begin{aligned}-\varphi(\Delta\varphi)\exp(2\lambda\psi^{-\nu})&=|\nabla_x\varphi|^2\exp(2\lambda\psi^{-\nu})\\ &+\sum_{i=1}^{n}\big[\psi_{x_ix_i}-(\nu+1)\psi^{-1}\psi_{x_i}^2\\ &\qquad-2\lambda\nu\psi^{-\nu-1}\psi_{x_i}^2\big]\lambda\nu\psi^{-\nu-1}\varphi^2\exp(2\lambda\psi^{-\nu})-B(\varphi),\end{aligned}$$

where

$$B(\varphi)=\lambda\nu\sum_{i=1}^{n}\big[\psi^{-\nu-1}\varphi^2\psi_{x_i}\exp(2\lambda\psi^{-\nu})+\varphi\varphi_{x_i}\exp(2\lambda\psi^{-\nu})\big]_{x_i}.$$

The next Lemma follows from Lemmas 2.6.2 and 2.6.3.

Lemma 2.6.4.. *Let* $\nu > \max\{6+\delta^{-2}, (n-1)\delta^{-2}+(n+1)\delta^4\}$, $\lambda \geq \lambda_0 > 0$, *where* λ_0 *is sufficiently large. Then the following inequality holds in* $\Omega(\gamma, \delta)$*:*

$$-3\lambda\nu\varphi(\Delta\varphi)\exp(2\lambda\psi^{-\nu}) + \psi^{\nu+1}(\Delta\varphi)^2\exp(2\lambda\psi^{-\nu})$$
$$\geq \lambda\nu|\nabla_x\varphi|^2\exp(2\lambda\psi^{-\nu}) + \frac{1}{2}\delta^4\lambda^3\nu^4\psi^{-2(\nu+1)}\exp(2\lambda\psi^{-\nu})$$
$$+ A(\varphi) + 3\lambda\nu B(\varphi).$$

Consider the following Cauchy problem for the system (2.6.1).

Problem 2.6.4. Given the conditions (2.6.6), (2.6.7), find the solution $(u^\pm, \varphi)$ of the system (2.6.1).

This problem was studied in Iordanskii (1961) in the case $t > 0$, $n = 1$, under the additional condition that $u(x,p,0) = u_1(x,p)$. It is evident that Problem 2.6.4 is ill posed in the classical sense.

Theorem 2.6.2. *Under the assumptions of Lemma 2.6.1, Problem 2.6.4 has at most one solution* $(u^\pm, \varphi)$ *such that* $\varphi \in \Phi$ *and* $u^\pm \in \mathfrak{M}$.

Proof. Let $(u_1^\pm, \varphi_1)$ and $(u_2^\pm, \varphi_2)$ be two solutions of Problem 2.6.2, both satisfying the same boundary conditions. Then, setting $u^\pm = u_2^\pm - u_1^\pm$, $\varphi = \varphi_2 - \varphi_1$, we conclude that the pair $(u^\pm, \varphi)$ satisfies the following system of equations ($x \in D$, $t > 0$, $p \in \mathbb{R}^n$):

$$l_\pm^{\varphi_1} u^\pm \mp e\langle\nabla_x\varphi, \nabla_p u_2^\pm\rangle = 0,$$
$$-\Delta_x\varphi = 4\pi\rho, \quad \rho = \int_{\mathbb{R}^n} (u^+ - u^-)\, dp$$

and the Cauchy data

$$u^\pm(0, 'x, p, t) = 0,$$
$$\varphi(0, 'x, t) = \varphi_{x_1}(0, 'x, t) = 0.$$

Let $\bar{u}^+(x,p,t) = u^+(x, -p, t)$. For $\bar{u}^+$, again, putting $u^+ = \bar{u}^+$ we obtain the system

$$l_\pm^{\varphi_1} u^\pm = \pm e\langle\nabla_x\varphi, \nabla_p u_2^\pm\rangle,$$
$$-\Delta_x\varphi = 4\pi\rho. \tag{2.6.17}$$

Then, using Lemma 2.6.1, we have

$$\lambda\nu\exp(2\lambda\psi^{-\nu})\frac{\delta_0}{m_\pm}(u^\pm)^2 + D^\pm(u^\pm)$$
$$\le [\langle\nabla_x\varphi, \nabla_p u_2^\pm\rangle]^2\psi^{\nu+1}\exp(2\lambda\psi^{-\nu}). \tag{2.6.18}$$

Under the assumptions of Lemma 2.6.4, the solution $(u^\pm, \varphi)$ of the system (2.6.17) satisfies the inequality

$$\lambda\nu|\nabla_x\varphi|^2\exp(2\lambda\psi^{-\nu}) + \psi^{-2(\nu+1)}\varphi^2\lambda^3\nu^4\exp(2\lambda\psi^{-\nu}) + B(\varphi)$$
$$\le 3\lambda\nu\exp(2\lambda\psi^{-\nu})\Big\{\varphi^2 + \operatorname{mes} G\int_{\mathbb{R}^n}[(u^+)^2 + (u^-)^2]\,dp\Big\}$$
$$+\psi^{\nu+1}\exp(2\lambda\psi^{-\nu})\operatorname{mes} G\int_{\mathbb{R}^n}[(u^+)^2 + (u^-)^2]\,dp. \tag{2.6.19}$$

This inequality can be obtained from Lemma 2.6.4, taking into account that $-\Delta_x\varphi = 4\pi\rho$. Multiplying inequality (2.6.18) by a number $k > 0$ such that $\dfrac{k\delta_0}{m_\pm} - 24\pi R^n > 1$ and adding it to (2.6.19), we have

$$\lambda\nu\exp(2\lambda\psi^{-\nu})\Big[\frac{k\delta_0}{m_\pm}(u^\pm)^2 + |\nabla_x\varphi|^2 + \psi^{-2(\nu+1)}\delta^4\lambda^2\nu^3\frac{\varphi^2}{2}\Big] + kD_\pm(u^\pm)$$
$$\le \exp(2\lambda\psi^{-\nu})\Big\{\big[\langle\nabla_x\varphi, \nabla_p u_2^\pm\rangle\big]^2 k\psi^{\nu+1} + 3\lambda\nu\varphi^2$$
$$-3\lambda\nu\operatorname{mes} G\int_{\mathbb{R}^n}[(u^+)^2 + (u^-)^2]\,dp\Big\}$$
$$+\psi^{\nu+1}\exp(2\lambda\psi^{-\nu})\operatorname{mes} G\int_{\mathbb{R}^n}[(u^+)^2 + (u^-)^2]\,dp.$$

We sum up these two inequalities written for each of the functions u^+ and u^- and integrate the resulting inequality over the domain $\Omega(\gamma, \delta)$, taking into account the condition $\dfrac{k\delta_0}{m_\pm} - 24\pi \operatorname{mes} G > 1$:

$$\begin{aligned}\lambda\nu \int_{\Omega(\gamma,\delta)} \Big\{ &\big[(u^+)^2 + (u^-)^2 + 2|\nabla_x\varphi|^2 \\ &+ \delta^4\lambda^2\nu^3\psi^{-2(\nu+1)}\varphi^2 - 6\varphi^2\big]\,\lambda\nu\exp(2\lambda\psi^{-\nu}) \\ &+ k(D(u^+) + D(u^-)) + 2B(\varphi) \\ &- \big[k\psi^{\nu+1}(\langle\nabla_x\varphi, \nabla_p u_2^\pm\rangle)^2 + k\psi^{\nu+1}(\langle\nabla_x\varphi, \nabla_p u_2^-\rangle)^2 \\ &+ 8\pi \operatorname{mes} G\psi^{\nu+1}((u^+)^2 + (u^-)^2)\big]\,\exp(2\lambda\psi^{-\nu})\Big\}\,dp\,dx\,dt \le 0.\end{aligned}$$

Passing to the limit as $\lambda \to \infty$ in the last inequality, one can easily prove that $u^\pm = 0$ and $\varphi = 0$ in $\Omega(\gamma, \delta)$.

Varying the point $x^0 = (0, x_2^0, \dots, x_n^0)$ in the plane $x_1 = 0$, we can prove that $u^\pm = \varphi = 0$ on the set of the form $\{(x,p,t) : (x,p,t) \in \Omega,\ 0 < x_1 < \delta\}$. Repeating the above arguments for the domain $\Omega_\delta = \Omega \cap \{x_1 > \delta\}$, we get $u^\pm = \varphi = 0$ on $\Omega \cap \{0 < x_1 < 2\delta\}$, and so on. This completes the proof of Theorem 2.6.2. □

Theorem 2.6.3. *In the system (2.6.1), let $g^\pm = G^\pm(x,p,t)\mu^\pm(x,p)$, where $G^\pm$ are known functions and $\mu^\pm$ are unknown; $G^\pm \neq 0$ in Ω. Then Problem 2.6.3 has at most one solution $(u^\pm, \varphi, \mu^\pm)$ such that $u^\pm \in \mathfrak{M}$, $\varphi \in \Phi$, $\mu \in C^1(D \times \mathbb{R}^n)$, and the functions φ, $\mu^\pm$ do not depend on t.*

Proof. Suppose that Problem 2.6.3 has two solutions, $(u_1^\pm, \varphi_1, \mu_1^\pm)$ and $(u_2^\pm, \varphi_2, \mu_2^\pm)$. For $u^\pm = u_2^\pm - u_1^\pm$, $\varphi = \varphi_2 - \varphi_1$, and $\mu^\pm = \mu_2^\pm - \mu_1^\pm$, we have

$$\begin{aligned} l_\pm^{\varphi_1} u^\pm &= \pm e\langle\nabla_x\varphi, \nabla_p u_2^\pm\rangle + G^\pm\mu^\pm, \\ -\Delta_x\varphi &= 4\pi\rho. \end{aligned} \tag{2.6.20}$$

Substituting $u^\pm = v^\pm G^\pm$ into (2.6.20), dividing the first equation by G^+ and the second by G^-, and differentiating them with respect to t, we obtain the following system for $w^\pm = v_t^\pm$ (using the condition $u^\pm(x,p,0) = 0$):

$$\begin{aligned} l_\pm^{\varphi_1} w^\pm + a^\pm w^\pm + a_t \int_0^t w^\pm(x,p,\tau)\,d\tau &= \pm\langle b_t^\pm, \nabla_x\varphi\rangle, \\ -\Delta_x\varphi = 4\pi \int_{\mathbb{R}^n} (w^+ - w^-)\,dp,& \end{aligned} \tag{2.6.21}$$

where $a^{\pm} = (G^{\pm})^{-1} l_{\pm}^{\varphi_1} G^{\pm}$, $b^{\pm} = \dfrac{1}{G^{\pm}} \nabla_p u_2$ and the following conditions hold:

$$w^{\pm}(0,'x,p,t) = 0, \quad \varphi(0,'x) = \varphi_{x_1}(0,'x) = 0. \tag{2.6.22}$$

As in Theorem 2.6.2, from (2.6.21), (2.6.22) it follows that $\varphi = w^{\pm} = 0$ in Ω because the term containing the integral of the form $\int_0^t w(x,p,\tau)\,d\tau$ satisfies the estimates of Lemma 2.6.4. □

2.7. SOME INVERSE AND DIRECT PROBLEMS FOR THE KINETIC EQUATION

This part of the section concerns the question of solution uniqueness for an inverse problem for the kinetic equation. Before considering this problem, we establish the uniqueness of solutions of two direct problems associated with it.

We introduce the following notation: Q is a domain in the real Euclidean space $\mathbb{R}^{2n+1}$ $(n \geq 1)$ of the variables (x,v,t), where $(x,v) \equiv p \in D$, and D is a bounded domain with boundary $\partial D \in C^1$, $t \in (0,T)$; $\bar{n}$ is the unit vector of outward normal to ∂Q, $\bar{n} = (n_x, n_v, n_t)$, $n_x = (n_{x_1}, \dots, n_{x_n})$, $n_v = (n_{v_1}, \dots, n_{v_n})$; $\Gamma = \partial D \times (0,T)$, $\Gamma_- = \{(p,t) : (p,t) \in \Gamma,\ \langle v, n_x\rangle + \langle f, n_v\rangle < 0\}$.

Consider the equation

$$lu + \beta u = \mu, \tag{2.7.1}$$

in the domain Q, where $lu \equiv u_t + \langle v, \nabla_x u\rangle + \langle f, \nabla_v u\rangle$, $\langle \cdot,\cdot\rangle$ is the scalar product in $\mathbb{R}^n$, and $f = (f_1, \dots, f_n)$.

Problem 2.7.1. Find a solution of equation (2.7.1) defined in the domain Q and satisfying the conditions

$$u = \varphi_1(p,t) \quad \text{on } \Gamma_-, \tag{2.7.2}$$

$$u(p,0) = \varphi_2(p) \tag{2.7.3}$$

Theorem 2.7.1. *Let $f \in C^1(\overline{Q})$, $\beta \in C(\overline{Q})$. Then Problem 2.7.1 has at most one solution in the space $C^1(Q)$.*

Proof. Assume $\mu = \varphi_1 = \varphi_2 = 0$ and set $z = u\exp(-\lambda t)$, where

$$2\lambda > \max_{(p,t)\in Q}\Big(\sum_{i=1}^{n}|f_{iv_i}| + 2|\beta|\Big).$$

Then the function z satisfies the relations

$$l_\lambda(z) \equiv l(z) + (\lambda+\beta)z = 0, \tag{2.7.1$'$}$$

$$z = 0 \quad \text{on } \Gamma_-, \tag{2.7.2$'$}$$

$$z(p,0) = 0. \tag{2.7.3$'$}$$

Multiplying equation (2.7.1′) by $2z$, we have

$$2l_\lambda(z)z = (z^2)_t + \sum_{i=1}^{n}[(v_i z^2)_{x_i} + (f_i z^2)_{v_i}] + \Big(2\lambda + 2\beta - \sum_{i=1}^{n} f_{iv_i}\Big)z^2 = 0.$$

Integrating this equality over Q and taking into account (2.7.2′) and (2.7.3′), we obtain

$$\int_Q\Big(2\lambda + 2\beta - \sum_{i=1}^{n} f_{iv_i}\Big)z^2\,dQ \le 0.$$

Hence $z = 0$ and therefore $u = 0$ in Q. □

Remark 2.7.1. In the stationary case of Problem 2.7.1, i.e., if $u_t = 0$ in (2.7.1), the condition (2.7.2) is sufficient for the uniqueness of solutions, provided that there exists i_0 such that $f_{i_0} > 0$ on D. To prove this, it suffices to introduce the function $z = u\exp(-\lambda v_{i_0})$ and repeat the arguments of the proof of Theorem 2.7.1.

Remark 2.7.2. It is worth comparing condition (2.7.2) with the boundary condition in the analogous problem for the transport equation where the values of φ_1 are specified at the points of Γ for which $\langle v, n_x\rangle \le 0$ (see Vladimirov, 1961; Case and Zweifel, 1967), and φ_1 is interpreted as the intensity of the external particle flux incident on the boundary of D at any time point $t \in (0,T)$.

Problem 2.7.2. Find a solution to equation (2.7.1) defined in Q and satisfying the conditions (2.7.2) and

$$u_t(p,0) = \varphi_2(p) \tag{2.7.4}$$

Theorem 2.7.2. *Assume that $f \in C^1(\overline{D})$ and $\beta \in C(\overline{D})$ are independent of t, and let $f_1 > 0$ on $\overline{D}$. Then Problem 2.7.2 has at most one solution in the space $C^2(Q)$.*

Proof. Let $\mu = \varphi_1 = \varphi_2 = 0$. If we differentiate equation (2.7.1) and condition (2.7.2) with respect to t and denote u_t by z, the function z will satisfy equation (2.7.1) and conditions (2.7.2), (2.7.3). Consequently, by Theorem 2.7.1, $z = u_t = 0$ in Q. Thus, we arrive at the homogeneous version of Problem 2.7.1 for the stationary equation (2.7.1), and Remark 2.7.1 implies that $u = 0$ in Q. □

In what follows we assume that the right-hand side of equation (2.7.1) has the form $\mu = g(p,t)\lambda(p)$, where $g(p,t)$ is some known function.

Consider the following inverse problem.

Problem 2.7.3. Find a pair of functions (u, λ) defined in Q which is a solution to Problem 2.7.2 with additional data

$$u(p,T) = \varphi_3(p) \tag{2.7.5}$$

Theorem 2.7.3. *Assume the following conditions to be satisfied:*

1. *the functions f, β, g are continuously differentiable in $\overline{Q}$ and f does not depend on t;*
2. *$f_1 > 0$ and $g \neq 0$ in $\overline{Q}$.*

Then Problem 2.7.3 has at most one solution $(u, \lambda) \in C^2(\overline{Q}) \times C^1(\overline{D})$.

Proof. Let $\varphi_1 = \varphi_2 = \varphi_3 = 0$. Introduce the new unknown function z such that $u = gz$. Then from (2.7.1) it follows that z satisfies the equation

$$lz + g^{-1}(lg + \beta)z = \lambda(p). \tag{2.7.6}$$

Differentiating (2.7.6) with respect to t, we get

$$\frac{\partial}{\partial t}(lz) = -(az_t + a_t z), \tag{2.7.7}$$

where $a = g^{-1}(lg + \beta)$.

With the auxiliary functions $\omega = z\exp(-\lambda v_1)$ and $\psi = -r(t+1)$, where $\lambda > 1$ and $r > \max\limits_{(p,t)\in\overline{Q}} (g_t/g)^2$ are constants, equation (2.7.7) takes the form

$$\frac{\partial}{\partial t}(l\omega) + \lambda f_1\omega_t + \psi\omega = \big[(\psi - a_t)z - az\big]e^{-\lambda v_1} \tag{2.7.8}$$

If we square both sides of (2.7.8), drop the positive term $[\partial(l\omega)/\partial t + \psi\omega]^2$, and apply the inequality $(a+b)^2 \le 2(a^2+b^2)$, we obtain the relation

$$\lambda^2 f_1^2\omega_t^2 + 2\lambda f_1\omega_t\Big[\frac{\partial}{\partial t}(l\omega) + \psi\omega\Big]$$
$$\le 2\big[(\psi - a_t)^2 z^2 + a^2 z_t^2\big]\exp(-2\lambda v_1). \tag{2.7.9}$$

Recalling that f_1 does not depend on t, it is easy to verify directly that

$$2\lambda f_1\omega_t\Big[\frac{\partial}{\partial t}(l\omega) + \psi\omega\Big]$$
$$= \lambda f_1(\omega_t^2 + \psi\omega)_t - \lambda f_1\psi_t\omega^2$$
$$+ \lambda\sum_{i=1}^{n}\big[(v_i f_1\omega_t^2)_{x_i} + (f_i f_1\omega_t^2)_{v_i}\big]$$
$$- \lambda\omega_t\sum_{i=1}^{n}\big[(f_1 f_i)_{v_i} + v_i f_{1x_i}\big] \tag{2.7.10}$$

By virtue of the identity (2.7.10), returning to the function z in the left-hand side of inequality (2.7.9), we have

$$\lambda f_1(z_t^2 + \psi z^2)_t\exp(2\lambda v_1)$$
$$+ \lambda\sum_{i=1}^{n}\big[(v_i f_1 z_t^2\exp(-2\lambda v_1))_{x_i} + (f_i f_1 z_t^2\exp(-2\lambda v_1))_{v_1}\big]$$
$$+ \Big\{\lambda^2 f_1^2 - \lambda\sum_{i=1}^{n}\big[(f_1 f_i)_{v_i} + v_i f_{1x_i}\big] - 2a^2\Big\} z_t^2\exp(-2\lambda v_1)$$
$$- \big[\lambda f_1\psi_t + 2(\psi - a_t)^2\big] z^2\exp(-2\lambda v_1) \le 0. \tag{2.7.11}$$

Let $\lambda > \max\{\lambda_1, \lambda_2\}$ $(\lambda > 1)$,

$$\max_{(p,t)\in\overline{Q}}\Big[\frac{1 - 2(\psi - a_t)^2}{-f_1\psi_t}\Big]^{1/2} < \max_{(p,t)\in\overline{Q}}\frac{1 + 2(\psi - a_t)^2}{f_1 r} = \lambda_1,$$

$$\lambda_2 = \max_{(p,t)\in Q} \frac{1}{f_1^2}\Big[\sum_1^n \big(|(f_1 f_i)_{v_i}| + |v_i f_{1x_i}|\big) + 2a^2\Big]. \tag{2.7.12}$$

Since $u = gz$, we obtain

$$z_t^2 + \psi z^2 = g^{-2}\big[u_t^2 + 2uu_t\, g^{-1} g_t + (\psi + g^{-2} g_t^2)u^2\big].$$

As follows from the conditions $\varphi_1 = \varphi_2 = 0$, (2.7.4), (2.7.5), and $r > \max\limits_{(p,t)\in\overline{Q}} (g^{-1}g_t)^2$, the foregoing equality implies

$$\begin{aligned} (z_t^2 + \psi z^2)\big|_{t=0} &= g^{-2}(\psi + g^{-2}g_t^2)u^2 \le 0, \\ (z_t^2 + \psi z^2)\big|_{t=T} &= g^{-2}u_t^2 \ge 0. \end{aligned} \tag{2.7.13}$$

Since $\varphi_1 = 0$, it is evident that

$$z_t = 0 \quad \text{on } \Gamma_- \tag{2.7.14}$$

Integrating inequality (2.7.11) over Q and taking into account (2.7.12), (2.7.13), (2.7.14), we get

$$\int_Q z^2 \exp(-2\lambda v_1)\, dQ \le 0,$$

i.e., $z = 0$ and therefore $u = 0$ in Q. Then (2.7.6) implies $\lambda = 0$, which completes the proof. □

Remark 2.7.3. The proof of Theorem 2.7.3 shows that the same result can be obtained for Problem 2.7.3 if equation (2.7.1) is replaced by the equation

$$lu + \beta u + \int_{Q_v} K(p, v', t)u(x, v', t)\, dv' = g(p,t)\lambda(p),$$

where Q_v is the orthogonal projection of Q onto the space $\mathbb{R}_v^n$, and the function K is continuously differentiable with respect to all its arguments.

Remark 2.7.4. The study of the uniqueness of solutions of the corresponding coefficient inverse problems for equation (2.7.1) is reduced to the study of this issue for Problem 2.7.3.

Chapter 3.

Evolutionary equations

This chapter deals with boundary value problems for special evolutionary equations associated with integral geometry and inverse problems for differential equations, particularly, for the kinetic equations. Below we present two problems of integral geometry associated with the problems that will be studied in this chapter.

Problem 1. (Bukhgeim, 1972b) Determine a function $\lambda(x, y)$ defined in $\mathbb{R}_+ = \{(x, y) : x \in \mathbb{R}^n,\ y > 0\}$ from the integrals of λ over a family of paraboloids $S(x, y)$, or more precisely, from the function

$$f(x, y) = \int_{S(x,y)} \lambda(\xi, \eta)\, \mathrm{d}\xi,$$

$$S(x, y) : \ \eta = y - \langle A(x - \xi), x - \xi\rangle.$$

Here A is a positive Hermitian matrix.

We restrict ourselves to the case where A is the unit matrix, i.e., $A = E$, $S(x, y) : \ \eta = y - |x - \xi|^2$, since the general case can be reduced to it by means of a linear change of variables. Then Problem 1 can be reformulated as follows: Determine $\lambda(x, y)$ from the function

$$f(x, y) = \int_{|x-\xi|\le\sqrt{y}} \lambda(\xi, y - |x - \xi|^2)\, \mathrm{d}\xi. \tag{1}$$

Let

$$u(x, y, t) = t^{-n/2} \int_{|x-\xi|\le\sqrt{y}} \lambda(\xi, y - t^{-1}|x - \xi|^2)\, \mathrm{d}\xi, \qquad t > 0. \tag{2}$$

Lemma. (Bukhgeim, 1972b) *Let* $\lambda \in C^\infty(\overline{\mathbb{R}}_+)$ *and* $\lambda(x,0) = \lambda'_y(x,0) = 0$. *Then* $u(x,y,t)$ *is infinitely differentiable for* $y > 0$, $t > 0$. *Moreover, the following relations are valid:*

$$L_1[u] = \Big(4\frac{\partial^2}{\partial t\,\partial y} - \Delta_x\Big)u = 0, \qquad (x,y) \in \mathbb{R}_+,\ t > 0, \tag{3}$$

$$u(x,y,1) = f(x,y), \quad u(x,0,t) = 0, \tag{4}$$

$$u(x,y,0) = \frac{\omega_n}{2}\int_0^y \lambda(x,s)(y-s)^{\frac{n}{2}-1}\,\mathrm{d}s, \tag{5}$$

where ω_n *is the area of the unit sphere in* $\mathbb{R}^n$.

Proof. It is easy to verify that

$$\Big(4\frac{\partial^2}{\partial t\,\partial y} - \Delta_x\Big)t^{-n/2}\lambda(\xi, y - t^{-1}|x-\xi|^2) \equiv 0$$

Since $\lambda(x,0) = \lambda'_y(x,0) = 0$, after the differentiation of the right-hand side of (2) the integrals over the boundary $|x-y| = \sqrt{yt}$ vanish. Therefore,

$$L_1[u] = \int_{|x-\xi|\le\sqrt{yt}} L_1\big[t^{-n/2}\lambda(\xi, y - t^{-1}|x-\xi|)\big]\,\mathrm{d}\xi = 0,$$

and equality (3) is proved. Equalities (4) follow immediately from (1), (2). To prove (5), we note that (2) can be rewritten as follows:

$$u(x,y,t) = \frac{1}{2}\int_{|\zeta|=1}\mathrm{d}\omega_\zeta\int_0^y \lambda(x + \sqrt{t(y-s)}\zeta, s)(y-s)^{\frac{n}{2}-1}\,\mathrm{d}s.$$

Putting $t = 0$ in the last equality, we obtain (5). The fact that $u(x,y,t)$ is infinitely differentiable is obvious. □

Thus, the study of the uniqueness of solutions of Problem 1 is reduced to the study of this issue for the Goursat problem ((4) – (5)) (see also Section 3.3).

If $\lambda = 0$ for $x_1 \le 0$, then, as follows from (2), u and u_{x_1} vanish when $x_1 = x_1^0 < 0$ for sufficiently small y and t, i.e., we arrive at Problem A (see Section 3.3).

Problem 2. Given the sum of integrals of $\lambda(x,y,t)$ along one edge and the opposite side of every regular three-sided pyramid with vertex angle equal to 90° and base lying in the plane $t = 0$, determine the function

$\lambda(x, y, t)$. It is assumed that the vertices of these pyramids fill some bounded domain in the half-space $t > 0$.

In Amirov (1977), Problem 2 is proved to be equivalent to a certain boundary value problem, namely, the "principal" problem which is indicated in Remark 3.3.3.

3.1. THE CAUCHY PROBLEM FOR AN INTEGRO-DIFFERENTIAL EQUATION

First, consider the Cauchy problem for the equation

$$\frac{\partial^m u}{\partial t^m} = a \frac{\partial^p}{\partial x^p} \int_0^y u(x, \tau, t)\, \mathrm{d}\tau \tag{3.1.1}$$

with the conditions

$$\left.\frac{\partial^r u}{\partial t^r}\right|_{t=0} = \varphi_r(x, y) \qquad r = 0, 1, 2, \ldots, m-1, \tag{3.1.2}$$

where $-\infty < x < \infty$, $0 \le y \le 1$, $0 \le t \le T$; m and p are natural numbers; $u(x, y, t)$ and $\varphi_r(x, y)$ are continuous complex-valued functions that are 2π-periodic in x; $a = \pm 1$; furthermore, $u(x, y, t)$ has continuous derivatives of order m with respect to t and of order p with respect to x.

A similar problem with $m = 1$, $p = 2$, and $a = \pm 1$ was studied in Bukhgeim (1972b).

Obviously, the functions $u(x, y, t)$ and $\varphi_2(x, y)$ can be expanded into the Fourier series with respect to x:

$$\begin{aligned} u(x, y, t) &= \sum_{-\infty}^{\infty} u_n(y, t) \mathrm{e}^{\mathrm{i}nx}, \\ \varphi_r(x, y) &= \sum_{-\infty}^{\infty} \varphi_{r,n}(y) \mathrm{e}^{\mathrm{i}nx}. \end{aligned} \tag{3.1.3}$$

Theorem 3.1.1. *Let one of the following conditions hold:*

1) $m = 1$, $p = 4k + 2$, $a = 1$; $k = 0, 1, 2, \ldots$

2) $m = 1$, $p = 4k$, $a = -1$; $k = 1, 2, 3, \ldots$

3) $m = 1$, $p = 2k + 1$, $a = \pm 1$; $k = 0, 1, 2, \ldots$

4) $p = 4k$, $a = 1$; $k = 1, 2, 3, \dots$

5) $p = 4k + 2$, $a = -1$; $k = 0, 1, 2, \dots$

Then, in the cases 1) and 2) the problem (3.1.1) – (3.1.2) is well posed in the classical sense in $L_2(D)$, *and in the cases 3) – 5) it is ill posed in the classical sense in* $L_2(D)$, *where* $D = [-\pi, \pi] \times [0, 1]$.

Proof. Consider the case 1) ($m = 1$, $p = 4k + 2$, $a = 1$; $k = 0, 1, 2, \dots$). Let $u(x, y, t) = u_1(x, y, t) + iu_2(x, y, t)$ be a solution to the problem (3.1.1) – (3.1.2). We set

$$g_r(t) = \int_{-\pi}^{\pi} \Big\{ \int_0^1 u_r^2(x, y, t)\, dy \Big\}\, dx;$$

$$v_r(x, y, t) = \int_0^y \frac{\partial^{2k+1} u_r(x, \tau, t)}{\partial x^{2k+1}}\, d\tau, \qquad r = 1, 2.$$

Since a is a real number ($a = 1$), the functions $u_r(x, y, t)$, $r = 1, 2$, satisfy equation (3.1.1). Then, since $u(x, y, t)$ is 2π-periodic in x, we have

$$\begin{aligned} g_r'(t) &= 2 \int_{-\pi}^{\pi} dx \Big\{ \int_0^1 u_r(x, y, t) [u_r(x, y, t)]_t' \, dy \Big\} \\ &= 2 \int_{-\pi}^{\pi} dx \Big\{ \int_0^1 \Big(\int_0^y \frac{\partial^{4k+2} u_r(x, \tau, t)}{\partial x^{4k+2}}\, d\tau \Big) u_r \, dy \Big\} \\ &= -2 \int_{-\pi}^{\pi} dx \Big\{ \int_0^1 \frac{\partial^{2k+1} u_r}{\partial x^{2k+1}} \Big(\frac{\partial^{2k+1} u_r(x, \tau, t)}{\partial x^{2k+1}}\, d\tau \Big)\, dy \Big\} \\ &= -2 \int_{-\pi}^{\pi} dx \Big\{ \int_0^1 (v_r)_y' v_r \, dy \Big\} = - \int_{-\pi}^{\pi} dx \Big\{ \int_0^1 \frac{\partial}{\partial y} (v_r^2)\, dy \Big\} \\ &= - \int_{-\pi}^{\pi} v_r^2(x, 1, t)\, dx \le 0, \end{aligned}$$

because $v_r(x, 0, t) = 0$.

Hence $g'(t) \le 0$ and therefore $g_r(t) \le g_r(0)$ for $t > 0$, which proves the theorem in the case 1). For the case 2) the proof is similar.

Consider the case 3) ($m = 1$, $p = 2k + 1$, $a = \pm 1$; $k = 0, 1, 2, \dots$). For definiteness we assume $a = 1$, $p = 4k+1$. Using the series expansions (3.1.3) for (3.1.1) – (3.1.2), we have

$$\begin{cases} \dfrac{\partial u_n}{\partial t} = (\mathrm{i}n)^p \displaystyle\int_0^y u_n(\tau, t)\,\mathrm{d}\tau = -\mathrm{i}^3 n^p \int_0^y u_n(\tau, t)\,\mathrm{d}\tau, \\ u_n\big|_{t=0} = \varphi_n(y), \qquad p = 4k + 1. \end{cases} \tag{3.1.4}$$

It is known (Krein, 1967) that the solution to this problem is

$$u_n = \{\exp(-\mathrm{i}^3 n^p t S)\}\varphi_n(y), \tag{3.1.5}$$

where

$$S\,\varphi_n(y) = \int_0^y \varphi_n(\tau)\,\mathrm{d}\tau, \quad \exp S = \sum_{k=0}^{\infty} \frac{S^k}{k!},$$

provided that problem (3.1.4) is considered in the space $L_2(0,1)$.

Using the formula

$$S^k \varphi = \frac{1}{(k-1)!} \int_0^y (y-\tau)^{k-1} \varphi(\tau)\,\mathrm{d}\tau,$$

we get

$$\{\exp(-\mathrm{i}^3 n^p t S)\}\varphi(y) = \varphi(y) + \sum_{k=1}^{\infty} \frac{(-1)^k (\mathrm{i}^3 n^p t)^k}{k!(k-1)!} \int_0^y (y-\tau)^{k-1} \varphi(\tau)\,\mathrm{d}\tau.$$

Setting $\varphi(y) \equiv 1$, we have

$$\begin{aligned} &\exp(-\mathrm{i}^3 n^p t S)\varphi(y) \\ &\quad = 1 + \sum_{k=1}^{\infty} \frac{(-1)^k (\mathrm{i}^3 n^p t y)^k}{(k!)^2} = \sum_{k=0}^{\infty} \frac{(-1)^k \left(\sqrt{\mathrm{i}^3 n^p t y}\right)^{2k}}{(k!)^2} \\ &\quad = I_0\left(2\sqrt{n^p t y}\right) = \operatorname{ber}\left(\sqrt{2n^p t y}\right) + \mathrm{i}\operatorname{bei}\left(2\sqrt{n^p t y}\right), \end{aligned} \tag{3.1.5'}$$

where

$$\begin{aligned} &\operatorname{ber} z = \frac{\mathrm{e}^{\alpha(z)}}{\sqrt{2\pi z}} \cos\beta(z), \quad \operatorname{bei} z = \frac{\mathrm{e}^{\alpha(z)}}{\sqrt{2\pi z}} \sin\beta(z) \\ &\alpha(z) \sim \frac{z}{\sqrt{2}} + \frac{1}{8\sqrt{2z}}, \qquad \beta(z) = \frac{z}{\sqrt{2}} - \frac{\pi}{8}, \end{aligned} \tag{3.1.6}$$

and $I_0(z)$ is the modified Bessel function of zero order.

Now we make the change of variables

$$2\sqrt{n^p t y} = z, \quad \mathrm{d}y = \frac{z\,\mathrm{d}z}{2n^p t}.$$

As a result, we have

$$\|I_0(2\sqrt{n^p t y})\|^2_{L_2(0,1)} = \int_0^1 |I_0(2\sqrt{n^p t y})|^2\,\mathrm{d}y$$

$$\geq \int_0^1 \mathrm{ber}^2(2\sqrt{n^p t y})\,\mathrm{d}y = \int_0^{2\sqrt{n^p t}} \mathrm{ber}^2 z\,\frac{z\,\mathrm{d}z}{2n^p t}.$$

Split this integral into three parts:

$$\frac{1}{2n^p t}\int_0^{2\sqrt{n^p t}} z\,\mathrm{ber}^2 z\,\mathrm{d}z$$

$$= \frac{1}{2n^p t}\int_0^{a_n} z\,\mathrm{ber}^2 z\,\mathrm{d}z + \int_{a_n}^{b_n} z\,\mathrm{ber}^2 z\,\mathrm{d}z + \int_{b_n}^{2\sqrt{n^p t}} z\,\mathrm{ber}^2 z\,\mathrm{d}z,$$

where

$$a_n = \Big\{\Big[\frac{2\sqrt{n^p t}}{\sqrt{2}\pi}\Big] - 1\Big\}\sqrt{2}\pi, \qquad b_n = a_n + \frac{\sqrt{2}\pi}{3},$$

and $[2\sqrt{n^p t}/\sqrt{2}\pi]$ denotes the integral part of $2\sqrt{n^p t}/\sqrt{2}\pi$. For sufficiently large z, from (3.1.6) we get

$$\mathrm{ber}\, z \approx \frac{\exp(z/\sqrt{2})}{\sqrt{2\pi z}}\cos\Big(\frac{z}{\sqrt{2}} - \frac{\pi}{8}\Big).$$

Then

$$\frac{1}{2n^p t}\int_{a_n}^{b_n} z\,\mathrm{ber}^2 z\,\mathrm{d}z$$

$$\approx \frac{1}{2n^p t}\int_{a_n}^{b_n} z\,\frac{\exp(\sqrt{2}z)}{2\pi z}\cos^2\Big(\frac{z}{\sqrt{2}} - \frac{\pi}{8}\Big)\,\mathrm{d}z$$

$$= \frac{1}{4\pi n^p t}\int_{a_n}^{b_n} \exp(\sqrt{2}z)\cos^2\Big(\frac{z}{\sqrt{2}} - \frac{\pi}{8}\Big)\,\mathrm{d}z$$

$$\geq \frac{b_n - a_n}{16\pi n^p t}\exp(\sqrt{2}a_n) = \gamma_n$$

because $|\cos(z/\sqrt{2} - \pi/8)| > 1/2$ for $z \in [a_n, b_n]$.

It is easy to see that $\gamma_n \to \infty$ as $n \to \infty$. This proves that the problem (3.1.1) – (3.1.2) is ill posed in the classical sense for $a = 1$, $p = 4k+1$, $m = 1$. To prove the same assertion for the cases $a = -1$, $p = 4k + 1$, $m = 1$, and $a = 1$, $p = 4k + 3$, $m = 1$, it suffices to consider the problem (3.1.4) for negative n (n is an integer).

We now turn to the case 4) ($p = 4k$, $a = 1$; $k = 1, 2, 3, \dots$). New unknown functions are introduced as follows:

$$\frac{\partial u_r}{\partial t} = u_{r+1}, \quad r = 0, 1, 2, \dots, m-1, \quad u_0 = u.$$

The problem (3.1.1) – (3.1.2) can be rewritten in the form

$$\begin{cases} \dfrac{\partial \hat{u}}{\partial t} = A\hat{u} \\ \hat{u}(x, y, 0) = \hat{\varphi}(x, y), \end{cases} \tag{3.1.7}$$

Here we set $\hat{u} = \{u_r(x, y, t)\}_0^{m-1}$, $\hat{\varphi}(x, y) = \{\varphi_r(x, y)\}_0^{m-1}$, $A = \|a_{k,s}\|_1^m$, $a_{k,k+1} = E$ ($k = 1, 2, \dots, m-1$; E is the identity operator in $L_2(D)$),

$$a_{m,1}\varphi(x, y) = a\frac{\partial^p}{\partial x^p}\int_0^y \varphi(x, \tau)\, d\tau,$$

and the other $a_{k,s} = \mathbf{0}$ ($\mathbf{0}$ is the zero operator in $L_2(D)$). We also put

$$\|\hat{u}\|_{L_2(0,1)}^2 = \sum_{r=1}^{m-1} \|u_r\|_{L_2(D)}^2.$$

Evidently, we can also take the Fourier series expansion in x of the functions u_r ($r = 0, 1, \dots, m-1$):

$$u_r(x, y, t) = \sum_{-\infty}^{\infty} u_{r,n}(y, t)\exp(inx). \tag{3.1.8}$$

Substituting the expansions (3.1.3) and (3.1.8) into (3.1.7), we obtain the problem

$$\begin{cases} \dfrac{\partial \hat{u}_n}{\partial t} = A_n\hat{u}_n \\ \hat{u}_n(y, 0) = \hat{\varphi}, \end{cases} \tag{3.1.7'}$$

where $\hat{u}_n = \{u_{r,n}\}_{r=0}^{m-1}$, $\hat{\varphi}_n = \{\varphi_{r,n}\}_{r=0}^{m-1}$, $A_n = \|a_{k,s}^{(n)}\|_1^m$, $a_{k,k+1}^{(n)} = E$ ($k = 1, 2, \ldots, m-1$; E is the identity operator in $L_2(0,1)$),

$$a_{m,1}^{(n)}\varphi = (in)^p \int_0^y \varphi(\tau)\,\mathrm{d}\tau = S_n$$

and the other $a_{k,s} = \mathbf{0}$ (here $\mathbf{0}$ is the zero operator in $L_2(0,1)$. Consider the Hilbert space H of the vector functions $\hat{\varphi}(y) = \{\varphi_r(y)\}_{r=0}^{m-1}$ with norm

$$\|\hat{\varphi}\|_H = \sum_{r=0}^{m-1} \|\varphi_r(y)\|^2_{L_2(0,1)}.$$

It is clear that the operator A_n is bounded in H:

$$\|A_n\| \le n^p. \tag{3.1.9}$$

The solution to the problem (3.1.7′) has the form $\hat{u} = \hat{\varphi}_n \exp(tA_n)$. Thus, it is easy to verify that the solution to the problem (3.1.7) is given by

$$\hat{u} = \sum_{-\infty}^{\infty} \exp(tA_n)\hat{\varphi}_n(y)e^{inx}, \tag{3.1.10}$$

where $\hat{\varphi}_n(y)$ are such that the series in (3.1.10) is convergent.

We shall now prove that $\|\exp(tA_n)\| \to \infty$ as $n \to \infty$. It suffices to prove that $\|B_n\| \to \infty$ as $n \to \infty$, where B_n is the first element of the first row of the matrix $\exp(tA_n)$. We can verify directly that for $1 \le q < m$

$$\begin{array}{ccccc} 1 & 2 & 3 & & q \end{array}$$

$$A_n^{km+q} = \begin{bmatrix} 0 & 0 & 0 & \cdots & 0 & S_n^k & 0 & 0 & \cdots & 0 \\ 0 & 0 & 0 & \cdots & 0 & 0 & S_n^k & 0 & \cdots & 0 \\ 0 & 0 & 0 & \cdots & 0 & 0 & 0 & S_n^k & \cdots & 0 \\ \cdot & \cdot & \cdot & \cdot & \cdot & \cdot & \cdot & \cdot & \cdot & \cdot \\ 0 & 0 & 0 & \cdots & 0 & 0 & 0 & 0 & \cdots & S_n^k \\ S_n^{k+1} & 0 & 0 & \cdots & 0 & 0 & 0 & 0 & \cdots & 0 \\ 0 & S_n^{k+1} & 0 & \cdots & 0 & 0 & 0 & 0 & \cdots & 0 \\ 0 & 0 & S_n^{k+1} & \cdots & 0 & 0 & 0 & 0 & \cdots & 0 \\ \cdot & \cdot & \cdot & \cdot & \cdot & \cdot & \cdot & \cdot & \cdot & \cdot \\ 0 & 0 & 0 & \cdots & S_n^{k+1} & 0 & 0 & 0 & \cdots & 0 \end{bmatrix}$$

and

$$A_n^{km} = \begin{bmatrix} S_n^k & 0 & 0 & \cdots & 0 \\ 0 & S_n^k & 0 & \cdots & 0 \\ 0 & 0 & S_n^k & \cdots & 0 \\ \cdot & \cdot & \cdot & \cdot & \cdot \\ 0 & 0 & 0 & \cdots & S_n^k \end{bmatrix}$$

where

$$S_n\varphi(y) = (in)^p \int_0^y \varphi(\tau)\,\mathrm{d}\tau,$$

Then $B_n = \sum\limits_{k=0}^{\infty} \frac{t^{km} S_n^k}{(km)!}$, and using (3.1.6) we get

$$B_n\varphi(y) = \varphi(y) + \sum_{k=1}^{\infty} \frac{n^{km} t^{km}}{(km)!(k-1)!} \int_0^y (y-\tau)^{k-1}\varphi(\tau)\,\mathrm{d}\tau.$$

Putting $\varphi(y) \equiv 1$, we have

$$B_n\varphi = \sum_{k=0}^{\infty} \frac{(n^p t^m)^k}{(km)!k!}.$$

Consequently, $\|B_n\| \to \infty$ as $n \to \infty$, which implies that the problem (3.1.1) - (3.1.2) is ill posed in the classical sense in $L_2(D)$.

Finally, consider the case 5) ($p = 4k+2$, $a = -1, k = 0, 1, 2, \dots$). In this case, the proof that the problem (3.1.1) - (3.1.2) is ill posed in the classical sense is similar to that in the case 4).

This completes the proof of Theorem 3.1.1. □

We now replace (3.1.1) in the problem (3.1.1) - (3.1.2) with the equality

$$\frac{\partial^m u}{\partial t^m} = a\Delta_x^p \int_0^y u(x,\tau,t)\,\mathrm{d}\tau, \tag{3.1.1'}$$

where $x = (x_1, x_2, \dots, x_l) \in \mathbb{R}^l$, Δ_x is the Laplacian of $u(x,y,t)$ with respect to x, the functions $\varphi_r(x,y)$ are periodic in x_j and satisfy the conditions of the problem (3.1.1) - (3.1.2).

From the proof of Theorem 3.1.1 it follows that its assertion, with the exception of the case 3), also holds for the problem (3.1.1′)–(3.1.2).

Remark 3.1.1. Instead of providing estimates of conditional stability for the problems (3.1.1) – (3.1.2) and (3.1.1′) – (3.1.2), we note that they can be obtained using the logarithmic convexity of the function $\varphi(t)$ introduced in Lemma 3.4.1. In the case where $m = 1$ and $p = 2k$, the estimates are a special case of Theorem 3.4.1.

3.2. THE PROBLEMS (3.1.1) – (3.1.2) FOR $M = 2K + 1$, $P = 1$ (THE CASE OF NONPERIODIC SOLUTIONS)

In this section, we study the following Cauchy problem:

$$\frac{\partial^{2k+1}}{\partial t^{2k+1}} u = a \frac{\partial}{\partial x} \int_0^y u(x, \tau, t)\, \mathrm{d}\tau, \tag{3.2.1}$$

with initial data

$$\left. \frac{\partial^r u}{\partial t^r} \right|_{t=0} = \varphi_r(x, y), \qquad r = 0, 1, 2, \dots, 2k, \tag{3.2.2}$$

where $0 \le x,\ y \le 1$, $t \in [0, \infty)$, $a = \pm 1$, the functions $u(x, y, t)$ and $\varphi_r(x, y)$ are continuous in all arguments, $u(x, y, t)$ has continuous derivative with respect to x and continuous derivative of order $(2k + 1)$ with respect to t. Moreover, for any point $(x, y) \in D = \{(x, y) : 0 \le x, y \le 1\}$, $\dfrac{\partial}{\partial x} u(x, y, t) \in L_1(0, \infty)$ and $u \in L_1(0, \infty) \cap L_2(0, \infty)$.

The following part of the section is devoted to the conditional well-posedness of the problem (3.2.1) – (3.2.2).

Applying the Laplace transform with respect to t to the problem (3.2.1) – (3.2.2), we obtain

$$\begin{aligned} &\lambda^{2k+1} \hat{u}(x, y, \lambda) - \varphi_{2k} - \lambda \varphi_{2k-1} - \dots \\ &\qquad - \lambda^{2k-1} \varphi_1 - \lambda^{2k} \varphi_0 = a \frac{\partial}{\partial x} \int_0^y \hat{u}(x, \tau, \lambda)\, \mathrm{d}\tau, \end{aligned}$$

where $\lambda = \gamma + ip$, $\gamma > 0$, and $\hat{u}(x, y, \lambda) = \displaystyle\int_0^\infty \mathrm{e}^{-\lambda t} u(x, y, t)\, \mathrm{d}t$.

Here we may pass to the limit as $\gamma \to \infty$ because $u \in L_1(0, \infty)$, $\partial u / \partial x \in L_1(0, \infty)$, and $u(x, y, t)$ is a continuous function of all its arguments which is continuously differentiable with respect to x.

We have

$$(ip)^{2k+1}\hat{u}(x,y,ip) - \varphi_{2k} - ip\,\varphi_{2k-1} - \dots - (ip)^{2k-1}\varphi_1 - (ip)^{2k}\varphi_0 = a\frac{\partial}{\partial x}\int_0^y \hat{u}(x,\tau,ip)\,d\tau. \quad (3.2.3)$$

Differentiating this equality, we obtain

$$(ip)^{2k+1}\frac{\partial}{\partial y}(u_1 + iu_2) - \frac{\partial\varphi_{2k}}{\partial y} - (ip)\frac{\partial\varphi_{2k-1}}{\partial y} - \dots - (ip)^{2k}\frac{\partial\varphi_0}{\partial y} = a\frac{\partial u_1}{\partial x} + ia\frac{\partial u_2}{\partial x}, \quad (3.2.4)$$

where $\hat{u}(x,y,ip) = u_1(x,y,p) + iu_2(x,y,p)$ is the Fourier transform of the function $u(x,y,t)$ with respect to t.

Then, by virtue of (3.2.3), from (3.2.4) we obtain the Cauchy problem for the system

$$\begin{cases} (-1)^k p^n \dfrac{\partial u_1}{\partial y} - a\dfrac{\partial u_2}{\partial x} = \dfrac{\partial g}{\partial y} \\ (-1)^{k+1} p^n \dfrac{\partial u_2}{\partial y} - a\dfrac{\partial u_1}{\partial x} = \dfrac{\partial h}{\partial y} \end{cases} \quad (3.2.5)$$

with initial data

$$u_1(x,0,p) = (-1)^k p^{-n} g(x,0), \qquad u_2(x,0,p) = (-1)^{k+1} p^{-n} h(x,0), \quad (3.2.6)$$

where $n = 2k+1$ and

$$g(x,y) = \sum_{r=1}^{k} (-1)^r p^{2r-1}\varphi_{2k-2r+1},$$

$$h(x,y) = \sum_{r=0}^{k} (-1)^r p^{2r}\varphi_{2k-2r}.$$

It is easy to verify that the system (3.2.5) is elliptic for any $p \in (-\infty,\infty)$. Therefore, the problem (3.2.5) – (3.2.6) has at most one solution in $C_1(D)$ (see Carleman, 1939). Thus, we have proved the following proposition.

Proposition 3.2.1. The problem (3.2.1) – (3.2.2) can have at most one solution.

Proposition 3.2.2. The problem (3.2.1) – (3.2.2) is ill posed in the classical sense in the space $L_2(\Pi)$, where $\Pi = D \times (0, \infty)$.

Proof. Setting $k = 2m$ and $a = 1$ for definiteness and making the change of variables $\tau = p^{-n}y$, we reduce the problem (3.2.5) – (3.2.6) to the Cauchy problem for the system

$$\begin{cases} \dfrac{\partial \tilde{u}_1}{\partial x} + \dfrac{\partial \tilde{u}_2}{\partial \tau} = -p^{-n}\dfrac{\partial h}{\partial \tau} \\ \dfrac{\partial \tilde{u}_1}{\partial \tau} + \dfrac{\partial \tilde{u}_2}{\partial x} = p^{-n}\dfrac{\partial g}{\partial \tau}, \qquad \tilde{u}_i = u_i(x, p^n\tau, p) \end{cases} \tag{3.2.7}$$

with initial data

$$\begin{aligned} \tilde{u}_1(x, 0, p) &= p^{-n}g(x, 0) \\ \tilde{u}_2(x, 0, p) &= -p^{-n}h(x, 0) \end{aligned} \tag{3.2.8}$$

The problem (3.2.7) – (3.2.8) is ill posed in the classical sense for any fixed $p \in (-\infty, \infty)$ because for the homogeneous version of (3.2.7) it is equivalent to the following well-known problem, which is ill posed in the classical sense (see Lavrent'ev, 1967): Determine an analytic function $\bar{f}(z)$ on $D(p) = \{(x, \tau) \,:\, 0 < x < p^{-n};\, 0 < \tau < p^{-n}$ for $p > 0$, and $p^{-n} < \tau < 0$ for $p < 0\}$, provided that $\bar{f}(z)$ is given on $\Gamma = \{(x, 0) \,:\, 0 < x < 1\}$, which is a part of the boundary of $D(p)$.

Then, recalling that $\hat{u} = u_1 + iu_2$ and using the Plancherel theorem, we establish that the problem (3.2.1) – (3.2.2) is ill-posed in the classical sense in the space $L_2(\Pi)$. □

We introduce the following notation:

$$\mathfrak{M} = \Big\{u(x, y, t) \,:\, u \in L_2(\Pi),\ |\hat{u}(x, y, ip)| \le \frac{M}{1 + |p|^n}\Big\},$$

$$\varphi(x, y) = \{\varphi_r(x, y)\}_0^{2k}, \quad \|\varphi\|_{C^q(D)} = \sum_{r=0}^{2k} \|\varphi_r\|_{C^q(D)}, \qquad q = 0, 1.$$

Theorem 3.2.1. *Assume that $u(x,y,t)$ is a solution to the problem (3.2.1) – (3.2.2), and let $u \in \mathfrak{M}$. Then the following stability estimate is valid:*

$$\|u\|_{L_2(\Pi)} \le \omega_1\big(\|\varphi\|_{C^1(D)}\big), \tag{3.2.9}$$

where $\omega_1(x) \in C[0,\infty)$ is an unknown nondecreasing function such that $\omega_1(0) = 0$.

Proof. Introducing $z = x + \mathrm{i}\tau$ and $\beta(z,p) = u_1(x, p^n\tau, p) + \mathrm{i}u_2(x, p^n\tau, p)$, for the solutions of the system (3.2.7) for any fixed p we get (see Vekua (1959), p. 42)

$$\beta(z,p) = \Phi(z,p) - \frac{1}{\pi}\iint_{D(p)} \frac{f(\zeta,p)}{\zeta - z}\,\mathrm{d}\xi\,\mathrm{d}\eta, \tag{3.2.10}$$

where

$$\zeta = \xi + \mathrm{i}\eta, \quad f(z,p) = \frac{1}{2}(G(z,p) + \mathrm{i}H(z,p)),$$
$$G(z,p) = p^{-n}\frac{\partial g(x,p^n,\tau)}{\partial \tau},$$
$$H(z,p) = p^{-n}\frac{\partial h(x,p^n,\tau)}{\partial \tau},$$

and $\Phi(z,p)$ is an analytic function in $D(p)$ for any $p \in (-\infty,\infty)$.

Since $\beta(z,p)$ is continuous in $\overline{D}(p)$ for any $p \in (-\infty,\infty)$, from (3.2.10) it follows that

$$\Phi(z,p)\big|_{\tau=0} = \beta(z,p)\big|_{\tau=0} + \frac{1}{\pi}\iint_{D(p)} \frac{f(\zeta,p)}{\zeta - z}\,\mathrm{d}\xi\,\mathrm{d}\eta\,\Big|_{\tau=0} \tag{3.2.11}$$

We now estimate $|\beta(z,p)|$ and $\displaystyle\frac{1}{\pi}\left|\iint_{D(p)} \frac{f(\zeta,p)}{\zeta - z}\,\mathrm{d}\xi\,\mathrm{d}\eta\right|$ for $\tau = 0$.

We have

$$|\beta(z,p)|\,\big|_{\tau=0} = (u_1^2(x,p^n\tau,p) + u_2^2(x,p^n\tau,p))^{1/2}\,\big|_{\tau=0}$$

$$\leq |p|^{-n}\Big(\Big(\sum_{r=1}^{k}(-1)^{r-1}|p|^{2r-1}|\varphi_{2k-2r+1}(x,0)|^2\Big)^{1/2}$$

$$+\Big(\sum_{r=0}^{k}(-1)^{r}|p|^{2r}|\varphi_{2k-2r}(x,0)|^2\Big)^{1/2}$$

$$\leq |p|^{-n}\Big\{\Big(\sum_{r=1}^{k}|p|^{4r-2}\Big)\Big(\sum_{r=1}^{k}|\varphi_{2k-2r+1}(x,0)|^2\Big)\Big\}^{1/2}$$

$$+|p|^{-n}\Big\{\Big(\sum_{r=0}^{k}|p|^{4r}\Big)\Big(\sum_{r=0}^{k}|\varphi_{2k-2r}(x,0)|^2\Big)\Big\}^{1/2}$$

$$\leq |p|^{-n}\Big(\sum_{r=0}^{2k}|\varphi_r(x,0)|^2\Big)^{1/2}\Big\{\Big(\sum_{r=1}^{k}|p|^{4r-2}\Big)^{1/2}+\Big(\sum_{r=0}^{k}|p|^{4r}\Big)^{1/2}\Big\}$$

$$\leq |p|^{-n}\Big(\sum_{r=0}^{2k}|\varphi_r(x,0)|^2\Big)\Big\{\Big(\sum_{r=1}^{k}|p|^{2r-1}\Big)+\Big(\sum_{r=0}^{k}|p|^{2r}\Big)\Big\}$$

$$\leq |p|^{-n}S(p)\,\|\varphi(x,p^n\tau)\|_{C(D(p))}, \tag{3.2.12}$$

where $S(p) = \sum_{r=1}^{2k}|p|^r$.

Next, for $\tau = 0$, similar calculations yield

$$\frac{1}{\pi}\left|\iint_{D(p)}\frac{f(\zeta,p)}{\zeta - z}\,d\xi\,d\eta\right| \leq K_0(p)S(p)\,|p|^{-n}\left\|\frac{\partial\varphi(x,p^n\tau)}{\partial\tau}\right\|_{C(D(p))}, \tag{3.2.13}$$

where

$$K_0(p) = \max_{z\in D(p)} K(z,p), \quad K(z,p) = \frac{1}{2}\iint_{D(p)}\frac{d\xi\,d\eta}{|\zeta - z|},$$

Note that $K_0(p)$ exists because $K(z,p)$ is a continuous function on $\overline{D}(p)$ for any fixed z (see Vekua (1959), p. 44).

Observing that $u \in \mathfrak{M}$, from (3.2.10) we have

$$|\Phi(z,p)| \leq |\beta(z,p)| + \frac{1}{\pi}\left|\iint_{D(p)} \frac{f(\zeta,p)}{\zeta - z}\, \mathrm{d}\xi\, \mathrm{d}\eta\right|$$

$$\leq \frac{M}{1+|p|^n} + K_0(p)S(p)\,|p|^{-n}\left\|\frac{\partial\varphi(x,p^n\tau)}{\partial\tau}\right\|_{C(D(p))}, \quad (3.2.14)$$

since inequality (3.2.13) is true for any point of $D(p)$, which is easy to verify.

Then, for any fixed $p \in (-\infty,\infty)$, from (3.2.11) – (3.2.14) we obtain the following estimate (analytic continuation, see Lavrent'ev, 1967):

$$\|\Phi(z,p)\|$$

$$\leq \left\{S(p)|p|^{-n}\left(\|\varphi(x,p^n\tau)\|_{C(D(p))} + K_0(p)\left\|\frac{\partial\varphi(x,p^n\tau)}{\partial\tau}\right\|_{C(D(p))}\right)\right\}^{1-\Psi(x,\tau,p)}$$

$$\times\left(\frac{M}{1+|p|^n} + S(p)\,|p|^n K_0(p)\left\|\frac{\partial\varphi(x,p^n\tau)}{\partial\tau}\right\|_{C(D(p))}\right)^{\Psi(x,\tau,p)} \quad (3.2.15)$$

where $\Psi(x,\tau,p)$ is a harmonic function with respect to (x,τ), and $0 \leq \Psi(x,\tau,p) \leq 1$. Using (3.2.13) and (3.2.15), from (3.2.10) we obtain

$$\|\beta(z,p)\|$$

$$\leq \left\{S(p)|p|^{-n}\left(\|\varphi(x,p^n\tau)\|_{C(D(p))} + K_0(p)\left\|\frac{\partial\varphi(x,p^n\tau)}{\partial\tau}\right\|_{C(D(p))}\right)\right\}^{1-\Psi(x,\tau,p)}$$

$$\times\left(\frac{M}{1+|p|^n} + S(p)\,|p|^{-n} K_0(p)\left\|\frac{\partial\varphi(x,p^n\tau)}{\partial\tau}\right\|_{C(D(p))}\right)^{\Psi(x,\tau,p)}$$

$$+ S(p)\,|p|^{-n} K_0(p)\left\|\frac{\partial\varphi(x,p^n\tau)}{\partial\tau}\right\|_{C(D(p))}$$

This inequality yields

$$|\hat{u}(x,y,\mathrm{i}p)| \leq \{S(p)\,|p|^{-n}(\|\varphi\|_{C(D)} + K_0(p)\|\varphi\|_{C^1(D)})\}^{1-\Psi_1}$$

$$\times\left(\frac{M}{1+|p|^n} + S(p)\,|p|^{-n}K_0(p)\|\varphi\|_{C^1(D)}\right)^{\Psi_1}$$

$$+ S(p)\,|p|^{-n}K_0(p)\|\varphi\|_{C^1(D)}, \quad (3.2.16)$$

where $\Psi_1(x,y,p) = \Psi(x,p^{-n}y,p)$.

Since $u \in \mathfrak{M}$, we get

$$\iint\limits_D \int_{-\infty}^{\infty} |\hat{u}(x,y,\mathrm{i}p)|^2 \, dx \, dy \, dp \le \iint\limits_D \int_{-1/\varepsilon}^{1/\varepsilon} |\hat{u}|^2 \, dx \, dy \, dp + \frac{2M^2 \varepsilon^{2n-1}}{2n-1}.$$

By (3.2.16) and the Plancherel theorem, the foregoing inequality implies

$$\begin{aligned}\|u\|_{L_2(\Pi)} \le \Big\{ \iint\limits_D \int_{-1/\varepsilon}^{1/\varepsilon} &\big(S(p)\,|p|^{-n}(\|\varphi\|_{C(D)} + K_0(p)\|\varphi\|_{C^1(D)})\big)^{2(1-\Psi_1)} \\ &\times \Big(\frac{M}{1+|p|^n} + S(p)\,|p|^{-n} K_0(p)\|\varphi\|_{C^1(D)}\Big)^{2\Psi_1} \, dx \, dy \, dp \Big\}^{1/2} \\ &+ \Big\{ \iint\limits_D \int_{-1/\varepsilon}^{1/\varepsilon} (S(p)\,|p|^{-n} K_0(p)\|\varphi\|_{C^1(D)})^2 \, dx \, dy \, dp \Big\}^{1/2} \\ &+ \Big(\frac{2}{2n-1}\Big)^{1/2} M \varepsilon^{n-1/2} \equiv F_1(\|\varphi\|_{C^1(D)}, \varepsilon).\end{aligned}$$

Setting

$$\omega_1(\|\varphi\|_{C^1(D)}) = \min_{\varepsilon} F_1(\|\varphi\|_{C^1(D)}, \varepsilon),$$

we finally arrive at inequality (3.2.9). □

Remark 3.2.1. For an even order of differentiation with respect to t in (3.2.1) instead of the odd order $(2k+1)$, the Cauchy data for this equation does not provide the uniqueness of solution of the problem (3.2.1) – (3.2.2). Indeed, assuming that the order of differentiation in (3.2.1) is two and $a = 1$ for the sake of simplicity, we have the problem (3.2.5) – (3.2.6) split into two problems

$$\frac{\partial u_r}{\partial x} + \frac{\partial u_r}{\partial y} = g_r(x,y,p), \tag{3.2.17}$$

$$u_r(x,0,p) = h_r(x,p), \tag{3.2.18}$$

where $r = 0, 1$, g_r and h_r are given functions determined from the functions $\varphi_r(x,y)$. Obviously, the data (3.2.18) is not sufficient to provide the uniqueness of solution for the problem (3.2.17) – (3.2.18).

3.3. BOUNDARY VALUE PROBLEMS

This section is devoted to boundary value problems for differential inequalities. They are associated with the problems of integral geometry presented at the beginning of Chapter 3. In particular, the results that will be proved here imply new uniqueness theorems for the solution of the Goursat problem for nonlinear hyperbolic equations.

1. Let Ω_α be a domain in $\mathbb{R}^{n+2}$ such that

$$\Omega_\alpha = \Big\{(x, y, t) \,:\, x \in \mathbb{R}^n,\, y > 0,\, t > 0,\, \alpha \geq 0,\, 0 < \alpha - y - t - \sum_{i=1}^{n} \frac{x_i^2}{2}\Big\}.$$

Consider the following differential inequality in Ω_α:

$$|u_{ty} + a^2 \Delta_x u| \leq C(|u| + |\nabla u|), \tag{3.3.1}$$

where C is a constant, $\nabla u = (\nabla_x u, u_y, u_t)$, and $a > 0$, with the conditions

$$u(x, y, 0) = u_0(x, y), \tag{3.3.2}$$

$$u(x, 0, t) = u_1(x, t). \tag{3.3.3}$$

Let $Lu \equiv a^{-1} u_{ty} + a \Delta_x u$ and $u_0 = u_1 = 0$. Then (3.3.1) implies

$$|Lu| \leq \frac{C}{a}(|u| + |\nabla u|).$$

We now introduce $u = v \exp(-\lambda \varphi)$ as a new unknown function, where $\lambda > 0$ and φ is a certain given function. Since we have

$$u_{yt} = (v_{yt} - \lambda \varphi_y v_t - \lambda \varphi_t v_y + \lambda^2 \varphi_t \varphi_y v - \lambda \varphi_{ty} v) \exp(-\lambda \varphi),$$

$$u_{x_i x_i} = (v_{x_i x_i} - 2\lambda \varphi_{x_i} v_{x_i} + (\lambda^2 \varphi_{x_i}^2 - \lambda \varphi_{x_i x_i}) v) \exp(-\lambda \varphi),$$

it follows that

$$(Lu)^2 \exp(2\lambda\varphi) = (a \Delta_x v + a^{-1} v_{yt} - 2\lambda a \sum_{i=1}^{n} \varphi_{x_i} v_{x_i}$$

$$- \lambda \varphi_t a^{-1} v_y - \lambda \varphi_y a^{-1} v_t + b_\varphi v)^2.$$

In this equality and throughout the rest of the section,

$$b_\varphi = \sum_{i=1}^{n} (\lambda^2 \varphi_{x_i}^2 - \lambda \varphi_{x_i x_i}) a + \lambda^2 a^{-1} \varphi_t \varphi_y - \lambda \varphi_{ty} a^{-1}.$$

Using the elementary inequality $(c+d)^2 \geq 2cd$, where

$$c = a\Delta_x v + a^{-1} v_{yt} + b_\varphi v - \lambda \psi a v,$$

$$d = \lambda \psi a v - 2\sum_{i=1}^{n} \lambda \varphi_{x_i} v_{x_i} a - \lambda \varphi_t a^{-1} v_y - \lambda \varphi_y v_t a^{-1},$$

we get

$$\begin{aligned}
&(Lu)^2 \exp(2\lambda\varphi) \\
&\quad \geq 2\lambda\psi a^2 v \Delta_x v + 2\lambda \psi v v_{ty} + 2\lambda \psi a b_\varphi v^2 - 2\lambda^2 \psi a^2 v^2 \\
&\quad - 4\lambda a^2 \Delta_x v \sum_{i=1}^{n} \varphi_{x_i} v_{x_i} - 4\lambda v_{yt} \sum_{i=1}^{n} \varphi_{x_i} v_{x_i} - 4\lambda a b_\varphi v \sum_{i=1}^{n} \varphi_{x_i} v_{x_i} \\
&\quad + 4\lambda a^2 \psi v \sum_{i=1}^{n} \varphi_{x_i} v_{x_i} - 2\lambda \varphi_t v_y \Delta_x v - 2\lambda a^{-2} \varphi_t v_y v_{yt} \\
&\quad - 2\lambda a^{-1} b_\varphi \varphi_t v_y v + 2\lambda \psi \varphi_t v v_y - 2\lambda \varphi_y v_t \Delta_x v \\
&\quad - 2\lambda \varphi_y a^{-2} v_t v_{yt} - 2\lambda \varphi_y a^{-1} b_\varphi v v_t + 2\lambda \varphi_y \psi v v_t, \qquad (3.3.4)
\end{aligned}$$

where $\psi \in C^2(\Omega_\alpha)$ is a given function.

To estimate the right-hand side of the last inequality, we divide it into several parts and calculate each part separately. For this purpose, each term of the sum will be transformed so that the resulting expression will contain only $v_{x_i}^2$, v_y^2, v_t^2, $v_{x_i} v_y$, $v_{x_i} v_t$, $v_y v_t$, v^2, and the divergent terms.

1) The first part:

$$\begin{aligned}
&2\lambda \psi a v (a\Delta_x v + a^{-1} v_{yt} + b_\varphi v - \lambda \psi a v) \\
&\quad = -2\lambda \psi a^2 |\nabla_x v|^2 - 2\lambda \psi v_y v_t \\
&\quad + \lambda v^2 \Big[\sum_{i=1}^{n} (\psi a^2)_{x_i x_i} + \psi_{ty} + 2\psi a b_\varphi - 2\lambda \psi^2 a^2\Big] \\
&\quad + 2\lambda \frac{\partial}{\partial y} (\psi v v_t) - \lambda \frac{\partial}{\partial y} (\psi_y v^2)_t
\end{aligned}$$

2) The second part:

$$4\lambda \sum_{i=1}^{n} a\varphi_{x_i} v_{x_i} (a\Delta_x v + a^{-1} v_{ty} + bv - \lambda\psi a v)$$

$$= 2\lambda \sum_{i,j=1}^{n} \Big[2(a^2\varphi_{x_i})_{x_j} v_{x_i} v_{x_j} - 2\frac{\partial}{\partial x_j}(a^2 \varphi_{x_i} v_{x_i} v_{x_j}) + 2\lambda (a^2 \varphi_{x_i} v_{x_j}^2)_{x_i} \Big]$$

$$+ 2\lambda \sum_{i=1}^{n} \Big\{ \varphi_{x_i t} v_{x_i} v_y + \varphi_{x_i y} v_{x_i} v_t - \varphi_{x_i x_i} v_y v_t - |\nabla_x v|^2 (a^2 \varphi_{x_i})_{x_i}$$

$$- [(ab_\varphi \varphi_{x_i} - \lambda\psi a^2 \varphi_{x_i})_{x_i}] v^2 + \frac{\partial}{\partial x_i}(\varphi_{x_i}(v_t v_y - ab_\varphi v^2 + \lambda\psi a^2 v^2))$$

$$- \frac{\partial}{\partial t}(\varphi_{x_i} v_{x_i} v_y) - (\varphi_{x_i} v_{x_i} v_t)_y \Big\}$$

3) The third part:

$$-2\lambda\varphi_t a^{-1} v_y (a\Delta_x v + a^{-1} v_{yt} + b_\varphi v - \lambda\psi a v)$$

$$= \lambda \sum_{i=1}^{n} (2\varphi_{x_i t} v_y v_{x_i} - \varphi_{ty} v_{x_i}^2) + \lambda (\varphi_t a^{-2})_t v_y^2$$

$$+ \lambda v^2 ((\varphi_t a^{-2} b_\varphi)_y - \lambda(\varphi_t \psi)_y) - 2\lambda \sum_{i=1}^{n} (\varphi_t v_y v_{x_i})_{x_i}$$

$$+ \frac{\partial}{\partial y}(\lambda\varphi_t |\nabla_x v|^2 - \varphi_t a^{-1} b_\varphi v^2 + \lambda\varphi_t \psi v^2) - \lambda\frac{\partial}{\partial t}(\varphi_t a^{-2} v_y^2)$$

4) The fourth part:

$$-2\lambda\varphi_y a^{-1} v_t (a\Delta_x v + a^{-1} v_{yt} + b_\varphi v - \lambda\psi a v)$$

$$= -\lambda\varphi_{yt} |\nabla_x v|^2 + \lambda(\varphi a^{-2})_y v_t^2 + 2\lambda \sum_{i=1}^{n} \varphi_{y x_i} v_t v_{x_i}$$

$$+ \lambda v^2 ((\varphi_y a^{-1} b_\varphi)_t - \lambda(\psi\varphi_y)_t) - 2\lambda \sum_{i=1}^{n} \frac{\partial}{\partial x_i}(v_t v_{x_i})$$

$$- \lambda\frac{\partial}{\partial y}(\varphi_y a^{-2} v_t^2) + \lambda\frac{\partial}{\partial t}(\varphi_y |\nabla_x v|^2 - \varphi_y a^{-1} b_\varphi v + \psi\varphi_y v^2)$$

Using identities 1) – 4), from (3.3.4) we obtain

$$
\begin{aligned}
&(Lu)^2 \exp(2\lambda\varphi) \\
&\quad \geq 4\lambda \sum_{i,j=1}^{n} (a^2\varphi_{x_i})_{x_j} v_{x_i} v_{x_j} - 2\lambda|\nabla_x v|^2 \Big(\varphi_{yt} + a^2\psi + \sum_{i=1}^{n}(a\varphi_{x_i})_{x_i}\Big) \\
&\quad + 4\lambda \sum_{i=1}^{n} (\varphi_{x_i t} v_{x_i} v_y + \varphi_{x_i y} v_{x_i} v_t) + \lambda(\varphi_t a^{-2})_t v_y^2 \\
&\quad + \lambda(\varphi_y a^{-2})_y v_t^2 - 2\lambda v_y v_t \Big(\psi + \sum_{i=1}^{n} \varphi_{x_i x_i}\Big) \\
&\quad + \lambda v^2 \Big\{ (\varphi_t a^{-1} b_\varphi)_y + (\varphi_y a^{-1} b_\varphi)_t + 2\psi a b_\varphi \\
&\qquad + \sum_{i=1}^{n} [(a b_\varphi \varphi_{x_i} - \lambda\psi a^2 \varphi_{x_i})_{x_i} + (\psi a^2)_{x_i x_i}] \\
&\qquad + \psi_{yt} - \lambda(\varphi_t\psi)_y - 2\lambda\psi^2 a^2 - \lambda(\psi\varphi_y)_t \Big\}, \\
&\quad + \lambda\, \mathrm{div}(V),
\end{aligned} \tag{3.3.5}
$$

where

$$
\begin{aligned}
\mathrm{div}(V) = \sum_{i=1}^{n} \frac{\partial}{\partial x_i} \Big(&\psi a^2 v v_{x_i} - (\psi a^2)_{x_i} v^2 - 2\sum_{j=1}^{n} a^2 (2\varphi_{x_j} v_{x_i} v_{x_j} - v_{x_i}^2) \\
&+ 2\varphi_{x_i} v_y v_t - 2\varphi_t v_y v_{x_i} - 2v_t v_{x_i} + 2v^2(\lambda\psi a - b_\varphi) a\varphi_{x_i} \Big) \\
+ \frac{\partial}{\partial y} \Big(&2\psi v v_t - 2\sum_{i=1}^{n} \varphi_{x_i} v_{x_i} v_t + \varphi_t |\nabla_x v|^2 \\
&- \varphi_y a^{-2} v_t^2 + \varphi_t v^2 (\psi - a^{-1} b_\varphi) \Big) \\
+ \frac{\partial}{\partial t} \Big(&\varphi_y |\nabla_x v|^2 - 2\sum_{i=1}^{n} \varphi_{x_i} v_{x_i} v_y - \varphi_t a^{-2} v_y^2 \\
&+ v^2 (\lambda\psi\varphi_y - \varphi_y a^{-1} b_\varphi - \psi_y) \Big).
\end{aligned}
$$

Here $a \in C^2(\Omega_\alpha)$, $a_y > 0$, $a_t > 0$, $a \geq a_0 > 0$. For a constant $\psi_0 > 0$, set $\psi = -\psi_0$ and $\varphi = \alpha - y - t - \frac{1}{2}\sum_{i=1}^{n} x_i^2$ in (3.3.5). We have

$$
\begin{aligned}
&(Lu)^2 \exp(2\lambda\varphi) \\
&\quad \geq 4\lambda \sum_{i,j=1}^{n} (a^2 x_i)_{x_j} v_{x_i} v_{x_j} - 2\lambda |\nabla_x v|^2 \Big(a^2\psi - \sum_{i=1}^{n} (a^2 x_i)_{x_i}\Big) \\
&\quad - \lambda(a^{-2})_t v_y^2 - \lambda(a^{-2})_y v_t^2 - 2\lambda v_t v_y(\psi + n) \\
&\quad + \lambda v^2 \Big[-\lambda^2(a^{-2})_y - \lambda^2(a^{-2})_t + 2\psi\lambda^2\Big(\sum_{i=1}^{n} a^2 x_i^2 + 1\Big) \\
&\qquad - \lambda^2 \sum_{i=1}^{n} \Big(x_i\Big(a^2 \sum_{j=1}^{n} x_j^2 + 1\Big)\Big)_{x_i} + 2\lambda\psi a^2 n \\
&\qquad + \lambda n \sum_{i=1}^{n} \Big((x_i a^2)_{x_i}(\psi - n) - \psi(a^2)_{x_i x_i}\frac{1}{\lambda}\Big) - 2\lambda^2\psi^2 a^2\Big]
\end{aligned}
\tag{3.3.6}
$$

Assume that $\psi_0 > 0$ is the smallest number with the property that the following quadratic form in $v_{x_1}, v_{x_2}, \ldots, v_{x_n}$ is positive definite:

$$
2\Big(a^2\psi_0 + \sum_{i=1}^{n} (x_i a^2)_{x_i}\Big) |\nabla_x v|^2 - 4 \sum_{i,j=1}^{n} (x_i a^2)_{x_j} v_{x_i} v_{x_j} \geq \beta\, |\nabla_x v|^2, \tag{3.3.7}
$$

where $\beta > 0$ is as small as desired. Such a number exists because $M \geq a \geq a_0 > 0$ and $|x|^2 < 1$ in the domain Ω_α.

Let $\nu_0 > 0$ be the smallest number satisfying the conditions below:

1. the following quadratic form in v_y, v_t is positive:

$$
2\nu_0 a^{-3}(v_y^2 + v_t^2) + 2v_t v_y(\psi_0 + n) \geq \beta(v_t^2 + v_y^2), \tag{3.3.8}
$$

2. the following inequality holds in Ω_α:

$$
4\nu_0 a^{-3} - 2\psi_0(a^2 + 1) - \sum_{i=1}^{n} \Big(x_i\Big(a^2 \sum_{j=1}^{n} x_j^2 + 1\Big)\Big)_{x_i} \geq 2\beta \tag{3.3.9}
$$

Suppose that

$$a_y > \nu_0, \quad a_t > \nu_0. \tag{3.3.10}$$

Evidently, from the conditions $a_y > 0$, $a_t > 0$, $M \geq a \geq a_0 > 0$ in $\overline{R}^+_{t,y}$ it follows that (3.3.10) can be obtained "in the small" using an appropriate change of variables y and t.

Note that if $a(x, y, t)$ is independent of x and $n > 2$, then $\psi_0 = 0$ and ν_0 depends only on $a(y, t)$ in (3.3.7). In the general case, ν_0 depends on a, a_{x_i} $(i = 1, 2, \dots, n)$, and α.

Let $\lambda_0 > 0$ be the smallest number satisfying the inequality

$$\lambda_0\beta - 2\psi_0 a^2(n + \psi_0) - \sum_{i=1}^{n}\left|(x_i a^2)_{x_i}(\psi_0 - n) - \frac{\psi_0}{\lambda}(a^2)_{x_i x_i}\right| \geq 0 \tag{3.3.11}$$

Then from (3.3.6) – (3.3.11) it follows that for $\lambda \geq \lambda_0 > 0$ we have

$$(Lu)^2 \exp(2\lambda\varphi) \geq \lambda\beta|\nabla_x v|^2 + \lambda^3\beta v^2 + \lambda \operatorname{div}(V). \tag{3.3.12}$$

We now return to the function $u(x, y, t)$ in the right-hand side of (3.3.12). For $\varphi = \alpha - y - t - \dfrac{1}{2}|x|^2$ we have

$$\begin{aligned}
&v = u\exp(\lambda\varphi), \quad v_\nu = (u_\nu + \lambda\varphi_\nu u)\exp(\lambda\varphi), \quad \nu = t, y, x_i,\\
&v_t^2 = (u_t^2 - \lambda^2\varphi_t^2 u^2)\exp(2\lambda\varphi) + \lambda(\varphi_t u^2\exp(2\lambda\varphi))_t,\\
&v_{x_i}^2 = (u_{x_i}^2 - \lambda^2\varphi_{x_i}^2 u^2)\exp(2\lambda\varphi) + \lambda(\varphi_{x_i} u^2\exp(2\lambda\varphi))_{x_i},\\
&v_y^2 = (u_y^2 - \lambda^2\varphi_y^2 u^2)\exp(2\lambda\varphi) + \lambda(\varphi_y u^2\exp(2\lambda\varphi))_y,
\end{aligned} \tag{3.3.13}$$

Consequently, for $\lambda \geq \lambda_0$, (3.3.12) implies

$$(Lu)^2\exp(2\lambda\varphi) \geq \lambda\beta|\nabla u|^2 - \lambda^3\beta\Big(1 + \sum_{i=1}^{n}\varphi_{x_i}^2\Big)u^2\exp(2\lambda\varphi) + \lambda\operatorname{div} U, \tag{3.3.14}$$

where

$$\begin{aligned}
\operatorname{div} U = \operatorname{div} u + \lambda\beta\Big(&\frac{\partial}{\partial y}(\varphi_y u^2\exp(2\lambda\varphi))\\
&+ \frac{\partial}{\partial t}(\varphi_t u^2\exp(2\lambda\varphi)) + \sum_{i=1}^{n}\frac{\partial}{\partial x_i}(\varphi_{x_i} u^2\exp(2\lambda\varphi))\Big),
\end{aligned}$$

and $\operatorname{div} u$ can be determined from the expression for $\operatorname{div}(V)$ in (3.3.5) by substituting the expressions for v, v_y, v_t, v_{x_i} in terms of u, u_y, u_t, u_{x_i} given by (3.3.13).

Furthermore, from (3.3.12) we obtain

$$(Lu)^2 \exp(2\lambda\varphi) \geq \lambda^3 \beta u^2 \exp(2\lambda\varphi) + \lambda \operatorname{div} U. \tag{3.3.15}$$

Multiplying inequality (3.3.15) by $(\alpha+2)$ and adding (3.3.14), for $|x|^2 \leq \alpha$ we get

$$\begin{aligned}(3+\alpha)(Lu)^2 \exp(2\lambda\varphi) &\geq \lambda\beta(|\nabla u|^2 + \lambda^2 u^2) \exp(2\lambda\varphi)\\ &\quad + \lambda(\operatorname{div} U + (2+\alpha)\operatorname{div} u)\end{aligned} \tag{3.3.16}$$

Using inequality (3.3.16), for $\lambda \geq \lambda_0$, from (3.3.1) we have

$$\begin{aligned}&2\frac{C^2}{a^2}(3+\alpha)(u^2 + |\nabla u|^2)\exp(2\lambda\varphi)\\ &\qquad \geq (3+\alpha)(Lu)^2 \exp(2\lambda\varphi)\\ &\qquad \geq \lambda\beta \exp(2\lambda\varphi) + \lambda^3\beta u^2 \exp(2\lambda\varphi) + \lambda(\operatorname{div} U + (2+\alpha)\operatorname{div} u)\end{aligned}$$

Hence, for sufficiently large $\lambda \geq \lambda_0$,

$$\frac{\beta u^2}{2} \leq \frac{\beta}{2} u^2 \exp(2\lambda\varphi) \leq -\frac{1}{\lambda^2}(\operatorname{div} U + (2+\alpha)\operatorname{div} u).$$

Passing to the limit as $\lambda \to +\infty$ in the last inequality and observing that $u_0 = u_1 = 0$, we see that $\frac{\beta}{2}\int_{\Omega_\alpha} u^2 \, d\Omega_\alpha \leq 0$. Hence $u = 0$ in Ω_α. Thus, we have proved the following lemma.

Lemma 3.3.1. *Assume that* $0 < \alpha < 1$, $a \in C^2(\overline{\Omega}_\alpha)$, *and* $a \geq a_0 > 0$, $a_y > \nu_0$, $a_t > \nu_0$ *in* $\overline{\Omega}_\alpha$, *where* ν_0 *is determined from (3.3.8), (3.3.9) and depends on* n, α, a, a_{x_i} $(i = 1, 2, \dots, n)$. *Let* $u(x, y, t) \in C^2(\Omega_\alpha) \cap C^1(\overline{\Omega}_\alpha)$ *be a solution of inequality (3.3.1) and satisfy the conditions (3.3.2) – (3.3.3) with* $u_0 = u_1 = 0$. *Then* $u \equiv 0$ *in* (Ω_α).

Remark 3.3.1. If the expression $u_{ty} + a^2\Delta_x u$ in (3.3.1) is reduced to the "usual" form (making the change of variables $\xi = t - y$, $\eta = y + t$), i.e.,

$$\bar{L}\tilde{u} \equiv -\tilde{u}_{\xi\xi} + \tilde{u}_{\eta\eta} + \tilde{a}^2 \Delta_x \tilde{u},$$

then the conditions $a_y > \nu_0$, $a_t > \nu_0$ are replaced by the condition $\tilde{a}_\eta > \nu_0$ if $\tilde{a}$ is independent of ξ. This condition means that in a medium with $\eta > 0$ the propagation speed of the disturbance described by the equation with the principal part $\bar{L}\tilde{u}$ increases with depth.

Remark 3.3.2. It is clear from the proof of Lemma 3.3.1 that if (3.3.1) is replaced by

$$|u_{yt} - a^2\Delta_x u| \le C(|u| + |\nabla u|), \tag{3.3.17}$$

then Lemma 3.3.1 will hold also for the problem (3.3.17), (3.3.2) - (3.3.3).

2. For $a = \text{const}$, the following problem can be studied: Find a solution $u(x, y, t)$ of the differential inequality

$$|u_{ty} - a^2 u_{xx}| \le C(|u| + |u_x|), \tag{3.3.18}$$

satisfying the conditions (3.3.2) - (3.3.3) and the condition

$$u(0, y, t) = u_2, \quad u_x(0, y, t) = u_3. \tag{3.3.19}$$

We shall call it Problem A for brevity. For Problem A, we shall prove a statement analogous to Lemma 3.3.1 in the domain $\widetilde{\Omega}_\alpha$:

$$\widetilde{\Omega}_\alpha = \Big\{(x, y, t) \,:\, y > 0,\, t > 0,\, x > 0, y + t < \frac{1}{2}(\alpha - x)^2\Big\},$$

where $\alpha > 0$ can be as large as desired. The proof is essentially similar to that of Lemma 3.3.1, with the only difference that $\varphi(x, y, t)$ is taken in the form $\varphi = \dfrac{1}{2}(\alpha - x)^2 - y - t$, $b_\varphi = \lambda^2(\varphi_x^2 a - a^{-1}) - \lambda a$, and $\psi = -1$.

Then the estimate (3.3.5) for Problem A will assume the form

$$(a^{-1}u_{xx} - u_{ty}a)^2 \exp(2\lambda\varphi) \\ \ge 4\lambda v_{xx}^2 + \lambda^3(1 + a^2\varphi_x^2)v^2 + \lambda\, \mathrm{div}(V), \tag{3.3.20}$$

In the same way as in the first part of this section, from (3.3.20) we derive

$$5(a^{-1}u_{xx} - u_{yt}a)^2 \exp(2\lambda\varphi) \\ \ge (4\lambda u_x^2 + \lambda^3 u^2)\exp(2\lambda\varphi) + \lambda\, \mathrm{div}\, U, \tag{3.3.21}$$

where

$$\operatorname{div} U = 4\lambda \frac{\partial}{\partial x}(\varphi_x u^2 \exp(2\lambda\varphi)) + \lambda \operatorname{div} u$$

If $u_i = 0$ $(i = 0, 1, 2, 3)$, then, passing to the limit as $\lambda \to +\infty$, from (3.3.18) and (3.3.21) we conclude that $u \equiv 0$ in $\widetilde{\Omega}_\alpha$.

Thus, we have proved the following lemma.

Lemma 3.3.2. *Suppose $u(x, y, t) \in C^2(\widetilde{\Omega}_\alpha) \cap C^1(\overline{\widetilde{\Omega}}_\alpha)$ is a solution to Problem A such that $u_i = 0$ $(i = 0, 1, 2, 3)$. Then $u \equiv 0$ in $\widetilde{\Omega}_\alpha$.*

Next, we shall establish the conditional stability estimate for Problem A. Let $u(x, y, t)$ be a solution to Problem A such that $u \in C^2(\widetilde{\Omega}_\alpha) \cap C^1(\overline{\widetilde{\Omega}}_\alpha)$ and

$$\|u\|_{C^2(\widetilde{\Omega}_\alpha)} \leq M, \quad \|u_i\|_{C(\Gamma_1)} \leq \varepsilon, \quad \|u_j\|_{C(\Gamma_2)} \leq \varepsilon, \tag{3.3.22}$$

$i = 2, 3$; $j = 0, 1$; $\Gamma_1 = \partial\widetilde{\Omega}_\alpha \cap \{x = 0\}$, $\Gamma_2 = \partial\widetilde{\Omega}_\alpha \cap \{y = 0\} \cup (\partial\Omega_\alpha \cap \{t = 0\})$. Then the following estimate holds:

$$\int_{\widetilde{\Omega}_\alpha} u^2 \, d\widetilde{\Omega}_\alpha \leq C\Big(\ln \frac{M}{\varepsilon}\Big)^{-2}. \tag{3.3.23}$$

Indeed, for sufficiently large λ, from (3.3.21) we obtain

$$\int_{\widetilde{\Omega}_\alpha} u^2 \exp(2\lambda\varphi) \, d\widetilde{\Omega}_\alpha \leq \frac{1}{\lambda^2} \int_{\partial\widetilde{\Omega}_\alpha} \langle \bar{\mathbf{n}}, U \rangle \, dS,$$

where $\bar{\mathbf{n}}$ is the outward normal to the surface $\partial\widetilde{\Omega}_\alpha$.

By virtue of (3.3.22), the foregoing inequality yields

$$\int_{\widetilde{\Omega}_\alpha} u^2 \, d\widetilde{\Omega}_\alpha \leq C_1 \lambda^{-2} (\varepsilon^2 \exp(\lambda\alpha^2) + M^2), \tag{3.3.24}$$

where $C_1 > 0$ is an absolute constant independent of λ, ε, M. Assuming that ε is sufficiently small, we take λ satisfying the condition $\varepsilon^2 \exp(\lambda\alpha^2) = M^2$, i.e., $\lambda = \dfrac{2}{\alpha^2} \ln \dfrac{M}{\varepsilon}$. Finally, (3.3.24) implies the desired estimate (3.3.23).

Remark 3.3.3. It is easily seen from the proof of Lemma 3.3.1 that its assertion still holds if (3.3.1) is replaced by the inequality

$$|u_{ty} \pm a^2 u_x| \leq C(|u| + |\nabla u|).$$

In the case where $a = \text{const}$, the assertion of Lemma 3.3.2 is true for the differential inequality which is obtained from (3.3.18) by substituting u_x for u_{xx}, condition (3.3.19) being replaced by the condition $u(0, y, t) = u_1$.

Remark 3.3.4. The results presented above in this section imply the uniqueness of solution in $C^2(\Omega_\alpha)$ of the Goursat problem for the nonlinear equation

$$u_{ty} \pm a^2 \Delta_x u = \mathcal{F}(x, y, t, u, u_{x_1}, \dots, u_{x_n}, u_t, u_y) \tag{3.3.25}$$

with conditions (3.3.2), (3.3.3), provided that $\mathcal{F}$ satisfies the Lipschitz condition for the arguments u, u_{x_i}, u_y, u_t $(i = 1, 2, \dots, n)$, and $a(x, y, t)$ satisfies the assumptions of Lemma 3.3.1.

Indeed, suppose that the problem (3.3.25), (3.3.2) – (3.3.3) has two solutions u_1 and u_2 in $C^2(\Omega_\alpha)$. Since $\mathcal{F}$ satisfies the Lipschitz condition for the corresponding arguments, $u = u_1 - u_2$ satisfies inequality (3.3.1) and conditions (3.3.2), (3.3.3) with $u_0 = u_1 = 0$. Then from Lemma 3.3.1 it follows that $u \equiv 0$ in Ω_α.

3.4. THE CAUCHY PROBLEM FOR AN EVOLUTIONARY EQUATION

1. Uniqueness and stability estimates.

Consider the following equation in a Hilbert space H:

$$\frac{\mathrm{d}u}{\mathrm{d}t} = A(t)u(t) + f(t), \tag{3.4.1}$$

where $A(t) = A_1(t)A_2$ for any $t \in (0, T)$, $A(t) = A_2 A_1(t)$, $A_1(t)(f(t))$ is a function with values in $[H](H)$ which is strongly measurable, bounded, and integrable in the sense of Bochner on $[0, T]$, $[H]$ is the Banach space of linear bounded operators on H, u(t) is a strongly differentiable function (for the definitions of strongly measurable and strongly differentiable functions, see Krein, 1967), A_2 is a self-adjoint operator, $A_2 = \int \lambda E(\mathrm{d}\lambda)$.

We consider the Cauchy problem for equation (3.4.1):

$$u(0) = u_0 \in D(A(t)). \tag{3.4.2}$$

Assume that $\sup_{t \in [0,T]} \|A_1(t)\| = 1$.

In this section, we study the conditional well-posedness of the problem (3.4.1) – (3.4.2). Previously, this subject was studied in Krein (1967), Krein and Prozorovskaya (1960) under different conditions imposed on the operators $A(t)$ for the homogeneous version of (3.4.1).

Lemma 3.4.1. *Let $\Phi(\lambda)$ be a nondecreasing positive function such that*

$$\int_{-\infty}^{\infty} |\lambda|^2 \exp(2\,|\lambda|\,T)\,\mathrm{d}\Phi(\lambda) < \infty.$$

Then the function $F(t) = \int_{-\infty}^{\infty} \exp(|\lambda|\,t)\,\mathrm{d}\Phi(\lambda)$ is logarithmically convex:

$$F(t) \le |F(0)|^{1-t/T}\,|F(T)|^{t/T}, \quad t \in [0,T].$$

Proof. We introduce the function

$$g(t) = \sum_{i=-\infty}^{\infty} \mathrm{e}^{\alpha_n t}\beta_n,$$

where the numbers $\alpha_n > 0$ and $\beta_n > 0$ satisfy the conditions

$$\sum_{i=-\infty}^{\infty} \alpha_n^r \mathrm{e}^{\alpha_n T}\beta_n < \infty, \quad r = 0, 1, 2.$$

Now consider the function $\varphi(t) = \ln g(t)$. Differentiating $\varphi(t)$ and using the Cauchy-Bunyakovsky inequality, we get

$$\begin{aligned}\varphi''(t) &= \frac{1}{g^2}(g''g - g'^2)\\ &= \frac{1}{g^2}\Big[\Big(\sum_{-\infty}^{\infty} \alpha_n^2 \mathrm{e}^{\alpha_n t}\beta_n\Big)\Big(\sum_{-\infty}^{\infty} \mathrm{e}^{\alpha_n t}\beta_n\Big)\\ &\qquad - \Big(\sum_{-\infty}^{\infty} \alpha_n \mathrm{e}^{\alpha_n t/2}\sqrt{\beta_n}\,\mathrm{e}^{\alpha_n t/2}\sqrt{\beta_n}\Big)^2\Big] \ge 0.\end{aligned}$$

This inequality implies that $\varphi(t)$ is lower convex, and therefore $\varphi(t)$ does not exceed the linear function on $[0,T]$ whose values at the ends of the interval coincide with those of $\varphi(t)$:

$$\varphi(t) \le \frac{T-t}{T}\varphi(0) + \frac{t}{T}\varphi(T).$$

Hence, after exponentiating, we have

$$|g(t)| \leq |g(0)|^{1-t/T} |g(T)|^{t/T}.$$

Application of this inequality to $F(t)$ completes the proof of the lemma. □

We introduce the following notation:

$$\mathfrak{M}_0 = \Big\{ u_0 \in D(A(t)) \, : \, \sum_{-\infty}^{\infty} \frac{T^n \| A_2^n u_0 \|}{n!} < \infty \Big\},$$

$$\mathfrak{M} = \Big\{ u_0 \in D(A(t)) \, : \, \sum_{-\infty}^{\infty} \frac{T^n \| A_2^n u_0 \|}{n!} \leq K \Big\},$$

$$\mathfrak{M}_i = \Big\{ u_0 \in D(A(t)) \, : \, \int_{-\infty}^{\infty} \exp(2\,|\lambda|\,T)(E(\mathrm{d}\lambda)u_0, u_0) \leq C_i^2, \\ \int_{-\infty}^{\infty} \exp(2\,|\lambda|\,T)|\lambda|^2 (E(\mathrm{d}\lambda)u_0, u_0) < \infty \Big\},$$

where $i = 1, 2$ and K, C_i are positive numbers.

Proposition 3.4.1. The problem (3.4.1) – (3.4.2) can have at most one solution.

Proof. Let $\alpha > 0$. Consider the operator

$$S = (\alpha + A_2^2)^{-1} = \int_{-\infty}^{\infty} \frac{E(\mathrm{d}\lambda)}{\alpha + \lambda^2} \tag{3.4.3}$$

Obviously, S is self-adjoint and $S \in [H]$, $\ker S = \{0\}$ (see Dunford and Schwartz, 1958).

Taking into account (3.4.2), we integrate (3.4.1) with respect to t and multiply both sides of the resulting equality by S to obtain

$$Su + (\mathcal{N}u)(t) = b(t), \tag{3.4.4}$$

where $\mathcal{N}(t) = \int_0^t SA(\tau)\,\mathrm{d}\tau$.

Since $A_1(t)$ commutes with A_2 and $\sup_{t\in[0,T]} \|A_1(t)\| \leq 1$, from (3.4.3) it follows that the operator $SA(t)$ is bounded and $\mathcal{N}(t)$ commutes with S. Therefore, by Proposition 2 of Bukhgeim (1972), equation (3.4.4) has at most one solution. □

It is easy to see that there exists a solution to the problem (3.4.1) – (3.4.2) if $u_0 \in \mathfrak{M}_0$ and $f(t)$ is a function with values in $\mathfrak{M}$ that is strongly measurable and integrable in the sense of Bochner.

Indeed, a solution to the following integral equation is a solution to the problem (3.4.1) – (3.4.2):

$$u(t) = b(t) + \int_0^t A(\tau)u(\tau)\,\mathrm{d}\tau.$$

Applying the method of successive approximations to this equation, we get

$$u(t) = b(t) + \int_0^t A_1(\tau)b(\tau)\,\mathrm{d}\tau \tag{3.4.5}$$
$$+\sum_{n=2}^{\infty}\int_0^t\int_0^{t_n}\cdots\int_0^{t_2} A(t_n)\cdots A(t_2)A(t_1)b(t_1)\,\mathrm{d}t_1\cdots\,\mathrm{d}t_n.$$

Here we have used the fact that the operators $A(t)$ are closed. $A(t)$ is closed because $A_1(t)$ is bounded and A_2 is self-adjoint.

Since $u_0 \in \mathfrak{M}$, $f(t) \in \mathfrak{M}$ for $t \in [0,T]$, and $f(t)$ is strongly measurable, it follows that the series (3.4.5) is convergent. One can verify directly that this series is a solution to the problem (3.4.1) – (3.4.2).

Theorem 3.4.1. *Assume that $u_0 \in \mathfrak{M}_1$ and $f(t)$ is a function with values in $\mathfrak{M}_2$ that is strongly measurable and integrable in the sense of Bochner. Then the following stability estimate holds for a solution to the problem (3.4.1) – (3.4.2):*

$$\|u(t)\| \le \|u_0\|^{1-t/T}C_1^{t/T} + (T)^{1/2}\Big(\int_0^T \|F(\tau)\|^2\,\mathrm{d}\tau\Big)^{(1-t/T)/2}C_2^{t/T}.$$

Proof. Since $A_1(t)$ commutes with A_2, from (3.4.5) it follows that

$$\begin{aligned} u(t) &= b(t) + \int_{-\infty}^{\infty} \lambda E(d\lambda) \int_0^t A_1(t_1) b(t_1)\, dt_1 \\ &+ \sum_{n=2}^{\infty} \int_{-\infty}^{\infty} \lambda^n E(d\lambda) \int_0^t \int_0^{t_n} \cdots \int_0^{t_2} A_1(t_n) \cdots A_1(t_1) b(t_1)\, dt_1 \cdots dt_n \\ &= b(t) + \int_{-\infty}^{\infty} E(d\lambda) \Big[\lambda \int_0^t A_1(\tau) b(\tau)\, d\tau \\ &\qquad + \sum_{n=2}^{\infty} \lambda^n \int_0^t \int_0^{t_n} \cdots \int_0^{t_2} A_1(t_n) \cdots A_1(t_1) b(t_1)\, dt_1 \cdots dt_n \Big] \end{aligned} \tag{3.4.6}$$

By the hypotheses of the theorem, the series (3.4.6) is convergent. Recalling that $b(t) = u_0 + \int_0^t f(\tau)\, d\tau$ and using the property of the projection operators

$$E(\lambda)E(\mu) = \begin{cases} E(\mu), & \lambda = \mu \\ 0, & \lambda \neq \mu \end{cases}$$

we establish that

$$\begin{aligned} \|u(t)\|^2 &\leq \int_{-\infty}^{\infty} \exp(2|\lambda| t)(E(d\lambda)u_0, u_0) \\ &+ \int_0^T \int_{-\infty}^{\infty} (E(d\lambda) f(s), f(s)) \exp(2|\lambda| t)\, ds \end{aligned} \tag{3.4.7}$$

Since $\sup_{t \in [0,T]} \|A_1(t)\| \leq 1$, we have

$$\Big\| \sum_{n=2}^{\infty} \lambda^n \int_0^t \int_0^{t_n} \cdots \int_0^{t_2} A_1(t_n) \cdots A_1(t_1)\, dt_1 \cdots dt_n + \int_0^t \lambda A_1(t_1)\, dt_1 + E \Big\|^2 \leq \exp(2|\lambda| t)$$

and

$$\Big(\int_0^t E(\mathrm{d}\lambda)f(s)\,\mathrm{d}s, \int_0^t E(\mathrm{d}\lambda)f(s)\,\mathrm{d}s\Big)$$
$$= \Big\|\int_0^t E(\mathrm{d}\lambda)f(s)\,\mathrm{d}s\Big\|^2 = \Big\|\int_0^t E(\mathrm{d}\lambda)f(s)\,\mathrm{d}s\Big\|\,\Big\|\int_0^t E(\mathrm{d}\lambda)f(s)\,\mathrm{d}s\Big\|$$
$$\leq \Big(\int_0^t \|E(\mathrm{d}\lambda)f(s)\|\,\mathrm{d}s\Big)^2 \leq T\int_0^T \|E(\mathrm{d}\lambda)f(s)\|^2\,\mathrm{d}s$$
$$= T\int_0^T (E(\mathrm{d}\lambda)f(s), f(s))\,\mathrm{d}s,$$

where $(\cdot\,,\cdot)$ denotes the scalar product in H.

Applying Lemma 3.4.1 to each term of the right-hand side of (3.4.9) and taking the square root of both sides of the resulting inequality, we arrive at the desired assertion. □

2. Regularization and the estimate of efficiency.

We set

$$R_\alpha = E + \int_{-1/\alpha}^{1/\alpha} \lambda E(\mathrm{d}\lambda)\int_0^t A_1(t_1)\,\mathrm{d}t_1 \tag{3.4.8}$$
$$+\sum_{n=2}^{\infty}\int_{-1/\alpha}^{1/\alpha} \lambda^n E(\mathrm{d}\lambda)\int_0^t\int_0^{t_n}\cdots\int_0^{t_2} A_1(t_n)\cdots A_1(t_1)\,\mathrm{d}t_1\cdots\,\mathrm{d}t_n,$$

and assume $u_0 \in \mathfrak{M}_1$, $u_\varepsilon \in D(A(t))$, $\|u_\varepsilon - u_0\| \leq \varepsilon$, $f(t) = 0$.

Theorem 3.4.2. *The family of linear operators $\{R_\alpha\}$ is a regularizing family for the problem (3.4.1) – (3.4.2) and the following estimate of efficiency holds for this regularizing family:*

$$\|R_\alpha u_\varepsilon - u(t)\| \leq \exp(t/\alpha)\varepsilon + C_1\exp(t-T)/\alpha.$$

Proof. It is easy to verify that $\{R_\alpha\}$ is a regularizing family for the problem (3.4.1) – (3.4.2) with $f(t) = 0$ and $\alpha \in (0,1]$. Indeed, for any $\alpha \in (0,1]$ the operator R_α is bounded and $R_\alpha u_0 \to u(t)$ as $\alpha \to 0$ for any $u_0 \in \mathfrak{M}_1$, since the series (3.4.6) is convergent for $u_0 \in \mathfrak{M}_1$.

Evidently, $\|R_\alpha\| \leq \exp(t/\alpha)$. From equalities (3.4.6), (3.4.8), and inequality (3.4.7) it follows that

$$\|R_\alpha u_0 - u(t)\|^2 \leq \int_{1/\alpha}^{\infty} \exp(2|\lambda|t)(E(\mathrm{d}\lambda)u_0, u_0) + \int_{-\infty}^{-1/\alpha} (E(\mathrm{d}\lambda)u_0, u_0)\exp(2|\lambda|t) \qquad (3.4.9)$$

It is clear that the sum on the right-hand side of this inequality attains its maximum value when

$$(E(\mathrm{d}\lambda)u_0, u_0) = C_1\,\chi(\lambda)\exp(-2|\lambda|t),$$

where $\chi(\lambda)$ is the characteristic function of the interval $\left[\frac{1}{\alpha}, \frac{1}{\alpha} + 1\right]$.

Then inequality (3.4.9) yields

$$\|R_\alpha u_0 - u(t)\| \leq C_1 \exp(t - T)/\alpha.$$

Consequently, if $\|u_0 - u_\varepsilon\| \leq \varepsilon$, then

$$\begin{aligned}\|R_\alpha u_\varepsilon - u(t)\| &\leq \|R_\alpha u_\varepsilon - R_\alpha u_0 + R_\alpha u_0 - u(t)\| \\ &\leq \|R_\alpha(u_\varepsilon - u_0)\| + \|R_\alpha u_0 - u_t\| \\ &\leq \varepsilon \exp(t/\alpha) + C_1 \exp(t - T)/\alpha.\end{aligned}$$

Therefore,

$$\|R_\alpha u_\varepsilon - u(t)\| \leq \varepsilon \exp(t/\alpha) + C_1 \exp(t - T)/\alpha.$$

□

Remark 3.4.1. All the above results obtained for the problem (3.4.1)–(3.4.2) remain valid if the operator A_2 is replaced by any closed operator B with dense domain of definition such that $B = VP$ in polar representation, where V is an isometric operator and P is a self-adjoint operator, $VP = PV$, and P commutes with $A_1(t)$ for any $t \in [0, T]$ (see Dunford and Schwartz, 1958).

3.5. INVERSE PROBLEM FOR AN EVOLUTIONARY EQUATION

Numerous works devoted to inverse problems are available nowadays (see Lavrent'ev, Romanov, and Shishatskii (1980) and the bibliography therein). Many of them concern the question of the uniqueness of solutions of the problems under study. In this section, we analyze the problem of solvability for one class of inverse problems for the right-hand side of an evolutionary equation in a real Hilbert space H. Iskenderov (1976) studied the uniqueness of solutions of inverse problems analogous to those considered in the present section.

By $C(0,T;H)$ we denote the space of functions $F:(0,T)\to H$ continuous on $[0,T]$, and by $L_2(0,T;H)$ the space of measurable functions $F(t)$ with values in H such that

$$\int_0^T \|F(t)\|^2\,dt = \|F\|^2_{L_2(0,T;H)} < \infty.$$

Throughout this section, $\|\cdot\|$ denotes the norm in H.

1. Consider the equation

$$u_{tt} + Au = f(t)g + F(t) \tag{3.5.1}$$

with constant linear operator with domain of definition $D(A)$ that is everywhere dense in H.

The function $u(t)$ is defined on $(0,T)$ and takes values in $D(A)$; $F(t)$ is defined on $(0,T)$, taking values in H; $f(t)$ is a given scalar function, and g is an unknown element of the space H.

The Cauchy data for equation (3.5.1) are given by

$$u(0) = u_0, \quad u_t(0) = u_1. \tag{3.5.2}$$

Problem 3.5.1. Determine the function $u(t)$ and the element $g \in H$ from equation (3.5.1) and the following information on the solution to the Cauchy problem (3.5.1) - (3.5.2):

$$u(T) = u_2. \tag{3.5.3}$$

Along with this problem, we consider another one:

Problem 3.5.1′. Find the pair $(u(t), g)$ from the equation

$$u_t + Au = f(t)g \tag{3.5.1'}$$

and the condition (3.5.3) for the solution to the Cauchy problem for this equation.

The methods of studying Problems 3.5.1 and 3.5.1′ are very similar. Here we shall confine ourselves to considering Problem 3.5.1. All the results obtained for this problem will be valid for Problem 3.5.1′.

Theorem 3.5.1. *Suppose A is a positive self-adjoint operator whose domain of definition $D(A)$ is dense in H, $f \in C^1(0,T)$, $f > 0$, $f_t > 0$.*

Then Problem 3.5.1 has at most one solution (u, g) such that $u(t)$, $u_t \in D(A^{1/2})$, $u_{tt} \in L_2(0,T;H)$ for any $t \in (0,T)$, and $\|u_{tt}\| < C$ in some fixed neighbourhoods of the points $t = 0$ and $t = T$, where $C > 0$ is a constant, $g \in H$, and $A^{1/2}$ is the square root of the operator A.

Proof. Assume that $F(t) = u_0 = u_1 = u_2 = 0$, where 0 is the zero element of H, and (u, g) is a solution to Problem 3.5.1 satisfying the conditions in the assertion of the theorem. From (3.5.1) we have

$$\frac{1}{f}(u_{tt} + Au) = g.$$

Taking the scalar product of this equality and u_t in H, we get

$$\left\langle \frac{1}{f}(u_{tt} + Au), u_t \right\rangle = \langle g, u_t \rangle \tag{3.5.4}$$

where $\langle \cdot, \cdot \rangle$ denotes the scalar product in H.

Since $f \in C^1(0,T)$, $u_{tt} \in L_2(0,T;H)$, it is easy to verify that

$$\left\langle \frac{1}{f} u_{tt}, u_t \right\rangle = \frac{1}{2}\left(\frac{1}{f}\|u_t\|^2\right)_t + \frac{f_t}{2f^2}\|u_t\|^2.$$

By the assumptions of Theorem 3.5.1, we have

$$\begin{aligned} \frac{1}{f}\langle Au, u_t \rangle &= \frac{1}{f}\langle A^{1/2}u, A^{1/2}u_t \rangle \\ &= \frac{1}{2}\frac{1}{f}(\|A^{1/2}u\|^2)_t = \frac{f_t}{2f^2}\|A^{1/2}u\| + \frac{1}{2}\left(\frac{1}{f}\langle Au, u \rangle\right)_t \end{aligned}$$

Therefore, (3.5.4) can be rewritten in the following form:

$$\frac{1}{f}\langle (u_{tt} + Au), u_t\rangle \tag{3.5.5}$$

$$= \frac{f_t}{2f^2}(\|u_t\|^2 + \|A^{1/2}u\|^2) + \frac{1}{2}\frac{\partial}{\partial t}\Big[\frac{1}{f}(\|u_t\|^2 + \langle Au, u\rangle)\Big] = \frac{\partial}{\partial t}(\langle g, u\rangle)$$

Using the condition that $u_0 = u_1 = u_2 = F = 0$ and $\|u_{tt}\| \le C$ in some neighbourhoods of $t = 0$ and $t = T$ (moreover, (3.5.1) implies $\|Au\| \le C_1$ for some C_1 in the same neighbourhoods), since (u, g) satisfies the assumptions of the theorem, integrating the last equality over $(0, T)$ we get

$$\int_0^T \frac{f_t}{2f^2}(\|u_t\|^2 + \|A^{1/2}u\|^2)\,\mathrm{d}t \le 0,$$

which implies $u_t = 0$. Hence $u = 0$ in $(0, T)$ because $u_0 = 0$. Finally, from (3.5.1) it follows that $g = 0$, which completes the proof. □

Theorem 3.5.2. *Let an operator A and a function f satisfy the assumptions of Theorem 3.5.1. Furthermore, assume that $F \in L_2(0,T;H)$ and A is weakly closed in $L_2(0,T;H)$ and has complete orthonormal system of eigenfunctions.*

Then for any u_0, u_1, $u_2 \in D(A)$ there exists a solution of Problem 3.5.1 with the following properties:

1. $u_{tt} \in L_2(0,T;H)$, $Au \in L_2(0,T;H)$, $g \in H$;
2. *$u(t)$ satisfies the conditions (3.5.2) – (3.5.4);*
3. *the pair (u, g) is a solution of equation (3.5.1).*

Proof. Consider the following equation, which is obtained from (3.5.1):

$$\frac{\partial}{\partial t}\Big[\frac{1}{f}(u_{tt} + Au)\Big] = \frac{\partial \mathcal{F}}{\partial t}, \tag{3.5.6}$$

whose solution satisfies conditions (3.5.2) – (3.5.4) with $\mathcal{F} = F/f$. Without loss of generality, hereafter we assume that $u_0 = u_1 = u_2 = 0$. First, consider the case $\mathcal{F}_t \in C(0,T;H)$.

By the problem (3.5.6), (3.5.2) – (3.5.4) we mean the problem of determining the function $u(t)$ from (3.5.6), provided that it satisfies conditions (3.5.2) – (3.5.4).

By the hypothesis of the theorem, A has complete orthonormal system of eigenfunctions $\{\omega_i\}$ and the corresponding eigenvalues $\{\lambda_i\}$, where λ_i are real and nonnegative. Let λ_i be indexed in increasing order and $\lambda_i > 0$ for all $i > s_0$.

The problem (3.5.6), (3.5.2) – (3.5.4) will be solved using the Galerkin method. We shall seek an approximate solution to (3.5.6) in the form $u_n(t) = \sum_{i=1}^{n} u_{ni}(t)\omega_i$, where $U_n = \{u_{n1}(t), \dots, u_{nn}(t)\}$ will be determined from

$$\Big\langle \frac{\partial}{\partial t}\Big\{\frac{1}{f}\Big[\sum_{i=1}^{n}(u_{ni\,tt}\omega_i + u_{ni}(t)A\omega_i\Big]\Big\}, \omega_i\Big\rangle = \langle \mathcal{F}_t, \omega_j\rangle, \tag{3.5.7}$$

$$u_{nj}(0) = 0, \quad \frac{\partial}{\partial t}u_{nj}(0) = 0, \quad u_{nj}(T) = 0. \tag{3.5.8}$$

Here $u_{nj}(t)$ are scalar functions on $(0, T)$.

The system (3.5.7) is a system of linear third order differential equations for the functions $u_{ni}(t)$, $1 \le i \le n$. We now prove that under the assumptions of Theorem 3.5.2 the problem (3.5.7) – (3.5.8) has a unique solution $U_n \in C^3(0, T)$ for any $\mathcal{F} \in C(0, T; H)$ and data (3.5.8). To this end, we shall prove that the homogeneous version of the problem (3.5.7) – (3.5.8) has only trivial solution (see Naimark, 1967).

Let $\widetilde{U}_n = (\tilde{u}_{n1}(t), \dots, \tilde{u}_{n1}(t)) \in C^3(0, T)$ be a solution to the homogeneous version of the problem. Multiply the jth equation of the system (3.5.7) by $-\tilde{u}_{nj}(t)$ and take the sum from 1 to n with respect to j. As a result, we get

$$-\Big\langle \frac{\partial}{\partial t}\Big[\frac{1}{f}(\tilde{u}_{n\,tt} + A\tilde{u}_n)\Big], \tilde{u}_n\Big\rangle = 0$$

where $\tilde{u}_n = \sum_{i=1}^{n} \tilde{u}_{ni}(t)\omega_i$.

Note that

$$-\Big\langle \frac{\partial}{\partial t}\Big[\frac{1}{f}(\tilde{u}_{n\,tt} + A\tilde{u}_n)\Big], \tilde{u}_n\Big\rangle \tag{3.5.9}$$

$$= -\frac{\partial}{\partial t}\Big[\Big\langle \frac{1}{f}(\tilde{u}_{n\,tt} + A\tilde{u}_n), \tilde{u}_n\Big\rangle\Big] + \Big\langle \frac{1}{f}(\tilde{u}_{n\,tt} + A\tilde{u}_n), \tilde{u}_{nt}\Big\rangle.$$

Using the first part of equality (3.5.5) (with $\tilde{u}_n = u$) and condition (3.5.8), we integrate (3.5.9) over $(0, T)$. We have $\int_0^T \|\tilde{u}_{nt}\|^2\, dt = 0$ and therefore

$\tilde{u}_{nt} = 0$. Since $\{\omega_i\}$ is linearly independent, from (3.5.8) it follows that $\tilde{u}_{ni} = 0$ for $1 \le i \le n$.

Thus, the problem (3.5.7) – (3.5.8) has a unique solution $U_n(t) \in C^3(0,T)$ for any right-hand side $\mathcal{F}_t \in C(0,T;H)$.

We shall now estimate the norm of $u_n(t) = \sum_{i=1}^{n} \omega_i u_{ni}(t)$ using the data of Problem 3.5.1, where $(u_{n1}(t), \dots, u_{nn}(t)) = U_n$ is a solution to the problem (3.5.7) – (3.5.8). Multiplying the jth equation of the system (3.5.7) by $-u_{nj}(t)$ and summing with respect to j, we obtain

$$-\Big\langle \frac{\partial}{\partial t}\Big[\frac{1}{f}(u_{n\,tt} + Au_n)\Big], u_n\Big\rangle = \langle \mathcal{F}_t, u_n\rangle. \tag{3.5.10}$$

From (3.5.9) with $\tilde{u}_n = u_n$, (3.5.5), condition (3.5.8) and the equality

$$-\langle \mathcal{F}_t, u_n\rangle = -\frac{\partial}{\partial t}\langle \mathcal{F}, u_n\rangle + \langle \mathcal{F}, u_{nt}\rangle$$

it follows that

$$|\langle \mathcal{F}, u_{nt}\rangle| \le \varepsilon^{-1}\|\mathcal{F}\| + \varepsilon\|u_{nt}\|^2,$$

where $\varepsilon < \min_{t\in[0,T]}(f_t/2f^2)$.

Integrating equality (3.5.10) over $(0,T)$, we get

$$\int_0^T \frac{f_t}{2f^2}(\|u_{nt}\|^2 + \|A^{1/2}u_n\|^2)\,\mathrm{d}t \le \int_0^T (\varepsilon^{-1}\|\mathcal{F}\|^2 + +\varepsilon\|u_{nt}\|^2)\,\mathrm{d}t$$

and therefore

$$\int_0^T (\|u_{nt}\|^2 + \|A^{1/2}u_n\|^2)\,\mathrm{d}t \le C. \tag{3.5.11}$$

Hereinafter, C stands for various constants that are independent of n.

To obtain estimates for $\|Au\|$, we multiply the jth equation of the system (3.5.7) by $-\lambda_j$ for $j \ge s_0$ and by zero for $j < s_0$. Then, summing from 1 to n with respect to j, we get

$$-\Big\langle \frac{\partial}{\partial t}\Big[\frac{1}{f}(u_{n\,tt} + Au_n)\Big], Au_n\Big\rangle = -\langle \mathcal{F}, Au_n\rangle \tag{3.5.12}$$

Transform the left-hand side of this equality as follows:

$$\Big\langle -\frac{\partial}{\partial t}\Big[\frac{1}{f}(u_{n\,tt} + Au_n)\Big], Au_n\Big\rangle \tag{3.5.13}$$

$$= -\frac{\partial}{\partial t}\Big\langle \frac{1}{f}(u_{n\,tt} + Au_n), Au_n\Big\rangle + \Big\langle \frac{1}{f}(u_{n\,tt} + Au_n), Au_{nt}\Big\rangle$$

It is easy to verify that

$$\Big\langle \frac{1}{f}(u_{n\,tt} + Au_n), Au_{nt} \Big\rangle \tag{3.5.14}$$

$$= \frac{1}{2}\frac{f_t}{f^2}(\|Au_n\|^2 + \|A^{1/2}u_{nt}\|^2) + \frac{\partial}{\partial t}\Big[\frac{1}{2f}(\|Au_n\|^2 + \|A^{1/2}u_{nt}\|^2)\Big]$$

Then, by virtue of equalities (3.5.13), (3.5.14), condition (3.5.8), and the inequality $|\langle \mathcal{F}_t, Au_n\rangle| \le \varepsilon^{-1}\|\mathcal{F}_t\|^2 + \varepsilon\|Au_n\|^2$, integrating (3.5.12) over $(0,T)$, we obtain

$$\int_0^T (\|Au_n\|^2 + \|A^{1/2}u_{nt}\|^2)\, dt \le C, \tag{3.5.15}$$

We now estimate $u_{n\,tt}$. From the system (3.5.7) it follows that

$$\frac{1}{f}\sum_{i=1}^n (u_{ni\,tt}\omega_i + u_{ni}A\omega_i) - \mathcal{F}(t) = c(t) + c_0, \tag{3.5.16}$$

where $c(t) \in C(0,T;H')$, H' is the orthogonal complement in H of the linear span of the elements $\omega_1, \dots, \omega_n$, and $c_0 = \dfrac{1}{f}(u_{n\,tt}(0) + \mathcal{F}(0))$.

Take the scalar product of equation (3.5.16) and $u_{n\,tt}$. We have

$$\frac{1}{f}(\|u_{n\,tt}\|^2 + \langle Au_n, u_{n\,tt}\rangle) = \langle c_0, u_{n\,tt}\rangle \tag{3.5.17}$$

The following equalities can be easily verified:

$$\frac{1}{f}\langle Au_n, u_{n\,tt}\rangle = \frac{\partial}{\partial t}\Big(\frac{1}{f}\langle Au_n, u_{nt}\rangle + \frac{f_t}{2f^2}\|A^{1/2}u_n\|^2\Big)$$
$$- \frac{1}{2}\Big(\frac{1}{f}\Big)_{tt}\|A^{1/2}u_n\|^2 - \frac{1}{f}\|A^{1/2}u_{nt}\|^2,$$

$$\langle c_0, u_{n\,tt}\rangle = \frac{\partial}{\partial t}\langle c_0, u_{nt}\rangle$$

By virtue of these relations and conditions (3.5.8), integrating equalities (3.5.17) over $(0,t)$, we have

$$\int_0^t \Big(\frac{1}{f(\tau)}\|u_{n\,\tau\tau}\|^2 - \frac{1}{2}\Big(\frac{1}{f}\Big)_{\tau\tau}\|A^{1/2}u_n\|^2$$
$$- \frac{1}{f(\tau)}\|A^{1/2}u_{n\tau}\|^2 - \langle F, u_{n\,\tau\tau}\rangle\Big)\, d\tau + \frac{1}{f}\langle Au_n, u_{nt}\rangle + \frac{f_t}{2f^2}\|A^{1/2}u_n\|^2$$
$$= \langle c_0, u_{nt}\rangle \equiv \frac{\partial}{\partial t}\langle c_0, u_n\rangle$$

Next, integrating the foregoing equality over $(0,T)$ and using the inequalities $|\langle F, u_{n\,\tau\tau}\rangle| \le \varepsilon^{-1}\|F\|^2 + \varepsilon\|u_{n\,\tau\tau}\|^2$, where $\varepsilon < \min_{t\in[0,T]} 1/f(t)$, we have

$$\int_0^T \mathrm{d}t\Big\{\int_0^t \Big[\frac{1}{f}\|u_{n\,\tau\tau}\|^2 - \frac{1}{2}\Big(\frac{1}{f}\Big)_{\tau\tau}\|A^{1/2}u_n\|^2 - \frac{1}{f}\|A^{1/2}u_{n\tau}\|^2\Big]\,\mathrm{d}\tau\Big\}$$

$$+\int_0^T \Big(\frac{1}{f}\langle Au_n, u_{nt}\rangle + \frac{f_t}{2f^2}\|A^{1/2}u_n\|^2\Big)\,\mathrm{d}t \le C \qquad (3.5.18)$$

Multiplying inequalities (3.5.11) by $T-\varepsilon_1$ and inequalities (3.5.15) by $T\varepsilon_2$, where $\varepsilon_1 > \dfrac{1}{2}\Big|\Big(\dfrac{1}{f}\Big)_{\tau\tau}\Big|$, $\varepsilon_2 > 1/f$, and summing the obtained inequalities and inequality (3.5.18), we arrive at the estimate

$$\int_0^T (T-t)\|u_{n\,\tau\tau}\|^2 \le C. \qquad (3.5.19)$$

Hence, for $0<\delta<T$, where δ is as small as desired, we have

$$\int_0^{T-\delta} \|u_{n\,\tau\tau}\|^2\,\mathrm{d}\tau \le C_\delta \qquad (3.5.19')$$

with C_δ independent of n.

Obviously, inequalities (3.5.11), (3.5.15), and (3.5.19) imply

$$\begin{aligned} &\|u_{nt}\|_{L_2(0,T;\,H)} \le C, \qquad \|u_n\|_{L_2(0,T;\,H)} \le C,\\ &\|Au\|_{L_2(0,T;\,H)} \le C, \qquad \|u_{n\,tt}\|_{L_2(0,T-\delta;\,H)} \le C. \end{aligned} \qquad (3.5.20)$$

Since $L_2(0,T;H)$ is a Hilbert space, the sequence $\{u_n(t)\}$ contains a subsequence that weakly converges to $u(t)$ in $L_2(0,T;H)$ (we again denote it by $\{u_n\}$). Furthermore, since $\|Au_n\|_{L_2(0,T;\,H)} \le C$, the sequence $\{Au_n\}$ also contains a subsequence, which we again denote by $\{Au_n\}$, that weakly converges to $\xi(t)$ in $L_2(0,T;H)$. The operator A being weakly closed, for any t we have $\xi(t)\in D(A)$ and $Au=\xi(t)\in L_2(0,T;H)$.

Similarly, from (3.5.20) it follows that $\{u_{nt}\}$ weakly converges to u_t in $L_2(0,T;H)$ and $\{u_{n\,tt}\}$ weakly converges to u_{tt} in $L_2(0,T-\delta;H)$ for $\delta>0$ as small as desired.

Next, multiply equalities (3.5.7) by a function $q(t)\in C_0^\infty(0,T)$, where $C_0^\infty(0,T)$ is the class of infinitely differentiable functions that are finite on $(0,T)$. Then transfer the derivative with respect to t onto $q(t)\omega_j$ and integrate the obtained equalities over $(0,T)$. As a result, for $n\ge j$ we have

$$\int_0^T \Big\langle \frac{1}{f}(u_{n\,tt}+Au_n), q(t)\omega_j)_t\Big\rangle\,\mathrm{d}t = \int_0^T \langle \mathcal{F}, (q(t)\omega_j)_t\rangle\,\mathrm{d}t.$$

Since the linear span of the functions of the form $q(t)\omega_j$, $1 \le j < \infty$, is everywhere dense in $L_2(0,T;H)$ for any $q(t) \in C_0^\infty(0,T)$ ($\{\omega_j\}$ is a complete system in H), passing to the limit as $n \to \infty$ in the above equality, we get

$$\int_0^T \langle \frac{1}{f}(u_{tt} + Au) - \frac{F}{f}, b_t \rangle \, dt = 0 \tag{3.5.21}$$

for any $b \in C_0^1(0,T;H)$, where $C_0^1(0,T;H)$ is the class of infinitely differentiable functions with values in H that are finite on $(0,T)$.

Write $g = \frac{1}{f}(u_{tt} + Au - F)$. Then, since $u_{tt} \in L_2(0,T-\delta;H)$, $Au \in L_2(0,T;H)$, and $F \in C(0,T;H)$, it follows that $g \in L_2(0,T-\delta;H)$ and (3.5.21) implies $\int_0^T \langle g, b_t\rangle \, dt = 0$ for any $b \in C_0^1(0,T;H)$. Therefore, g does not depend on t. By the definition of g and the fact that $Au \in L_2(0,T;H)$, it follows that $u_{tt} \in L_2(0,T;H)$.

We now prove that $u(t)$ satisfies condition (3.5.3). Let $\eta \in C^1(0,T;H)$ and $\eta(0) = \eta_t(0) = 0$. Since $\{u_n\}$ and $\{u_{nt}\}$ weakly converge in $L_2(0,T;H)$ to u and u_t, respectively, we see that

$$\begin{gathered} \int_0^T \langle u_n, \eta\rangle \, dt \to \int_0^T \langle u, \eta\rangle \, dt \\ \int_0^T \langle u_{nt}, \eta\rangle \, dt \to \int_0^T \langle u_t, \eta\rangle \, dt \end{gathered} \tag{3.5.22}$$

as $n \to \infty$. On the other hand, by the conditions $u_t \in L_2(0,T;H)$ and $0 = u_n(0) = u_n(T) = \eta(0) = \eta_t(0)$,

$$\begin{gathered} \int_0^T \langle u_{nt}, \eta\rangle \, dt = -\int_0^T \langle u_n, \eta_t\rangle \, dt \\ \int_0^T \langle u_t, \eta\rangle \, dt = -\int_0^T \langle u, \eta_t\rangle \, dt + \langle u(T), \eta(T)\rangle \end{gathered} \tag{3.5.23}$$

From (3.5.21) and (3.5.23) it follows that $\langle u(T), \eta(T)\rangle = 0$. Since $\eta(t)$ is arbitrary, it follows that $u(T) = 0$.

The proof of the fact that $u(t)$ satisfies condition (3.5.2) is similar.

Thus, Theorem 3.5.2 is proved in the case $F_t \in C(0,T;H)$. Let $F_t \in L_2(0,T;H)$. Clearly, there exists a sequence of functions $F_N(t)$ such that $\frac{\partial}{\partial t} F_N \in C(0,T;H)$ and $\frac{\partial F_N}{\partial t} \to F_t$ as $N \to \infty$ in $L_2(0,T;H)$. As was proved before, for any F_N there exists a pair (u_N, g_N) satisfying the assumptions of Theorem 3.5.2. The estimates (3.5.20) hold for the function $u_N(t)$; moreover, $\|g_N\| \le C$, $\|u_{N\,tt}\|_{L_2(0,T;H)} \le C$, and

$$u_{N\,tt} + Au_N = f(t)g_N + F_N. \tag{3.5.24}$$

From the *a priori* estimates for u_N and g_N it follows that there are subsequences of the sequences $\{u_N\}$ and $\{g_N\}$ for which there exist weak limits as $N \to \infty$ in $L_2(0,T;H)$ in (3.5.24). We shall use the same designations $\{u_N\}$ and $\{g_N\}$ for the subsequences. Evidently, the weak limit (u, g) satisfies equation (3.5.1). Using the *a priori* estimates (3.5.20), it is easy to prove that $u(t)$ satisfies conditions (3.5.2) – (3.5.4). This completes the proof of Theorem 3.5.2. □

2. Similarly to the way we studied Problem 3.5.1, in this part of the section we study the inverse problem for the following equation given in the space H:

$$u_{tt} - Au = g + F(t), \tag{3.5.25}$$

where A is a given positive self-adjoint operator (as in the previous part), $g \in H$ is unknown, $F(t) \in L_2(0,T;H)$ is a given function, $t \in (0,T)$.

Problem 3.5.2. Determine the functions $u(t)$ and $g \in H$ from equation (3.5.25) and the following conditions:

$$u(0) = u_0, \quad u_t(0) = u_1, \tag{3.5.26}$$

$$u(T) = u_2. \tag{3.5.27}$$

As in the previous part, without loss of generality, in Problem 3.5.2 we may assume that $u_0 = u_1 = u_2 = 0$.

Theorem 3.5.3. *Let A be a constant positive self-adjoint operator with domain of definition $D(A)$ which is dense in H. Then Problem 3.5.2 has at most one solution with the properties specified in Theorem 3.5.1.*

Proof. In equation (3.5.25), we introduce the new unknown function $u(t) = (t + \alpha)v(t)$, where $\alpha > 0$. Hence, for $v(t)$ we have the equation

$$(t + \alpha)(v_{tt} - Av) + 2v_t = g + F(t), \tag{3.5.28}$$

and the data

$$v(0) = v_t(0) = 0, \tag{3.5.29}$$

$$v(T) = 0. \tag{3.5.30}$$

Let $F = u_i = 0$ ($i = 0, 1, 2$) and let (v, g) be a solution to the problem (3.5.28) – (3.5.30) with the properties specified in Theorem 3.5.1. Take the scalar product of equation (3.5.28) and $v(t)$ in H. We have

$$\langle (t+\alpha)(v_{tt} - Av) + 2v_t, v_t \rangle = \langle g, v_t \rangle$$

Furthermore, the following equality holds:

$$\begin{aligned} &\langle (t+\alpha)(v_{tt} - Av) + 2v_t, v_t \rangle \\ &= \frac{1}{2}(3\|v_t\|^2 + \langle Av, v \rangle) + \frac{1}{2}\frac{\partial}{\partial t}[\,(t+\alpha)(\|v_t\|^2 - \langle Av, v \rangle)\,]. \end{aligned}$$

Using the arguments similar to those in the proof of Theorem 3.5.1, from this equality we deduce $v = 0$, $g = 0$. □

Theorem 3.5.4. *Let the operator A satisfy the hypothesis of Theorem 3.5.2 and $F_t \in L_2(0, T; H)$. Then for any $u_i \in D(A)$, $i = 1, 2, 3$, there exists a solution (u, g) of Problem 3.5.2 with the properties specified in Theorem 3.5.2.*

Proof. From (3.5.8) we obtain

$$\frac{\partial}{\partial t}[(t+\alpha)(v_{tt} - Av) + 2v_t] = \frac{\partial F}{\partial t}. \tag{3.5.31}$$

First, we assume that $F_t \in C(0, T; H)$.

The problem of finding the function $v(t)$ from (3.5.31) that satisfies conditions (3.5.29), (3.5.30) will be called the problem (3.5.31), (3.5.29), (3.5.30). This problem will be studied with the use of the Galerkin method.

We shall seek a function of the form $u_n(t) = \sum\limits_{i=1}^{n} u_{ni}(t)\omega_i$ such that

$$\Big\langle \frac{\partial}{\partial t}[(t+\alpha)(u_{n\,tt} - Au_n) + 2u_{nt}],\, \omega_j \Big\rangle = \Big\langle \frac{\partial F}{\partial t}, \omega_j \Big\rangle, \tag{3.5.32}$$

$$u_{nj}(0) = u_{nj}(T) = \frac{\partial}{\partial t} u_{nj}(0) = 0, \tag{3.5.33}$$

where $\{\omega_j\}$ is the system of eigenvectors of the operator A that correspond to the eigenvalues $\{\lambda_j\}$.

The homogeneous version of (3.5.32) – (3.5.33) has only trivial solution in $C^3(0,T)$. Indeed, let $\widetilde{U}_n = (\tilde{u}_{n1}, \dots, \tilde{u}_{nn}) \in C^3(0,T)$ be a solution to the homogeneous version of (3.5.32) – (3.5.33). Multiply by $-\tilde{u}_{nj}$ the jth equation of the system (3.5.32) written for $\widetilde{U}_n$ and sum the obtained equalities from 1 to n with respect to j. It is easy to verify that

$$\Big\langle -\frac{\partial}{\partial t}[(t+\alpha)(\tilde{u}_{n\,tt} - A\tilde{u}_n) + 2\tilde{u}_{nt}],\, \tilde{u}_n\Big\rangle$$

$$= \frac{1}{2}(3\|\tilde{u}_{nt}\|^2 + \langle A\tilde{u}_n, \tilde{u}_n\rangle) + \frac{1}{2}\frac{\partial}{\partial t}[(t+\alpha)(\|\tilde{u}_{nt}\|^2 - \langle A\tilde{u}_n, \tilde{u}_n\rangle$$

$$- 2\langle \tilde{u}_{n\,tt} - A\tilde{u}_n, \tilde{u}_n\rangle) - 4\langle \tilde{u}_{nt}, \tilde{u}_n\rangle] = 0, \tag{3.5.34}$$

where

$$\tilde{u}_n = \sum_{j=1}^{n} \tilde{u}_{nj}\omega_j.$$

Integrating equality (3.5.34) over $(0,T)$, we obtain

$$\int_0^T (3\|\tilde{u}_{nt}\|^2 + \langle A\tilde{u}_n, \tilde{u}_n\rangle)\, dt = 0,$$

or $\partial\tilde{u}_{nj}/\partial t = 0$ for $1 \le j \le n$. Finally, from (3.5.33) it follows that $\tilde{u}_{nj} = 0$.

Therefore, the problem (3.5.32) – (3.5.33) has a unique solution $U_n(t) = (u_{n1}, \dots, u_{nn}) \in C^3(0,T)$ for any right-hand side $F_t \in C(0,T;H)$.

***A priori* estimates.**

a) Multiplying (3.5.32) by $-u_{nj}(t)$ and summing with respect to j, we have

$$-\Big\langle \frac{\partial}{\partial t}[(t+\alpha)(u_{n\,tt} - Au_n) + 2u_{nt}],\, u_n\Big\rangle = -\Big\langle \frac{\partial F}{\partial t}, u_n\Big\rangle.$$

Integrating this equality over $(0,T)$, by virtue of (3.5.34) (with u_n replaced by $\tilde{u}_n$), we get

$$\frac{1}{2}\int_0^T (3\|\tilde{u}_{nt}\|^2 + \langle A\tilde{u}_n, \tilde{u}_n\rangle)\, dt \le -\int_0^T \Big\langle \frac{\partial F}{\partial t}, u_n\Big\rangle\, dt.$$

From this inequality, one can easily obtain the estimate

$$\int_0^T (\|u_{nt}\|^2 + \langle Au_n, u_n\rangle) \le C, \tag{3.5.35}$$

where C stands for different constants independent of n.

b) In (3.5.32), ω_j can be replaced by $\frac{1}{\lambda}A\omega_j$. Then (3.5.32) implies

$$-\Big\langle \frac{\partial}{\partial t}[(t+\alpha)(u_{n\,tt}-Au_n)+2u_{nt}],\ A\omega_j\Big\rangle = -\Big\langle \frac{\partial F}{\partial t}, A\omega_j\Big\rangle.$$

Consequently,

$$-\Big\langle \frac{\partial}{\partial t}[(t+\alpha)(u_{n\,tt}-Au_n)+2u_{nt}],\ Au_n\Big\rangle = -\Big\langle \frac{\partial F}{\partial t}, Au_n\Big\rangle. \tag{3.5.36}$$

The following equality is obvious:

$$\begin{aligned}-\Big\langle \frac{\partial}{\partial t}&[(t+\alpha)(u_{n\,tt}-Au_n)+2u_{nt}],\ Au_n\Big\rangle\\ &= \frac{1}{2}(\|Au_n\|^2+3\langle Au_{nt},u_{nt}\rangle)\\ &\quad -\frac{\partial}{\partial t}\Big[\langle (t+\alpha)(u_{n\,tt}-Au_n)+2u_{nt},\ Au_n\rangle\\ &\qquad -\frac{1}{2}(t+\alpha)\,\|A^{1/2}u_{nt}\|^2+\frac{1}{2}(t+\alpha)\,\|Au_n\|^2\Big].\end{aligned}$$

Hence, integrating (3.5.36) over $(0,T)$, we have

$$\frac{1}{2}\int_0^T(\|Au_n\|^2+3\langle Au_{nt},u_{nt}\rangle)\,dt \le \int_0^T\Big\langle \frac{\partial F}{\partial t}, Au_n\Big\rangle\,dt,$$

which implies the inequality

$$\int_0^T(\|Au_n\|^2+3\langle Au_{nt},u_{nt}\rangle)\,dt \le C. \tag{3.5.37}$$

c) From the system (3.5.32) it follows that

$$(t+\alpha)(u_{n\,tt}-Au_n)+2u_{nt} = c_0+c(t)+F(t), \tag{3.5.38}$$

where $c_0=\alpha u_{n\,tt}(0)-F(0)$ and $c(t)$ is an unknown function satisfying the condition $\langle c(t),\omega_j\rangle=0$ for $1\le j\le n$.

Take the scalar product of (3.5.18) and $u_{n\,tt}$ in H:

$$\langle (t+\alpha)(u_{n\,tt}-Au_n)+2u_{nt}-F,\ u_{n\,tt}\rangle = \langle c_0, u_{n\,tt}\rangle \tag{3.5.39}$$

It is obvious that

$$\langle (t+\alpha)(u_{n\,tt} - Au_n) + 2u_{nt},\, u_{n\,tt}\rangle$$
$$= (t+\alpha)(\|u_{n\,tt}\|^2 + \langle Au_{nt}, u_{nt}\rangle)$$
$$+ \frac{\partial}{\partial t}\Big(\frac{1}{2}\|A^{1/2}u_n\|^2 - (t+\alpha)\langle Au_n, u_{nt}\rangle + \|u_{nt}\|^2\Big).$$

By virtue of the foregoing equality and condition (3.5.33), integrating (3.5.38) over $(0,t)$ we get

$$\int_0^t (\tau + t)(\|u_{n\,\tau\tau}\|^2 + \langle Au_{n\tau}, u_{n\tau}\rangle)\, d\tau$$
$$+ \frac{1}{2}\|A^{1/2}u_n\|^2 - (t+\alpha)\langle Au_n, u_{nt}\rangle + \|u_{nt}\|^2$$
$$\le \int_0^t \langle F, u_{n\,\tau\tau}\rangle\, d\tau + \frac{\partial}{\partial t}\langle c_0, u_n\rangle.$$

Next, we integrate this relation over $(0,T)$ using the inequalities

$$\int_0^t \langle F, u_{n\,\tau\tau}\rangle\, d\tau \le \frac{2}{\alpha}\int_0^t \|F\|^2\, d\tau + \frac{\alpha}{2}\int_0^t \|u_{n\,\tau\tau}\|^2\, d\tau$$

and sum up the result with inequalities (3.5.35) and (3.5.37) to establish the estimate

$$\int_0^T dt\Big[\int_0^t (\|u_{n\,\tau\tau}\|^2 + \langle Au_{n\tau}, u_{n\tau}\rangle)\, d\tau\Big] \le C.$$

Hence, for $0 < \delta < T$ we have

$$\int_0^{T-\delta} (\|u_{n\,tt}\|^2 + \langle Au_{nt}, u_{nt}\rangle)\, dt \le C_\delta, \tag{3.5.40}$$

where δ is as small as desired and C_δ is a constant that depends on δ and the data of Problem 3.5.2.

In exactly the same manner as in the proof of Theorem 3.5.2, using inequalities (3.5.35), (3.5.37), and (3.5.40), we can prove that the problem (3.5.28) – (3.5.30) has a solution $(v(t), g)$ such that $v_t t \in L_2(0,T;H)$, $Av \in L_2(0,T;H)$, and $g \in H$. Then the pair (u,g), where $u = (t+\alpha)v$, is the desired one. This completes the proof of Theorem 3.5.4. □

Remark 3.5.1. Theorems analogous to Theorems 3.5.1–3.5.4 hold for the problems obtained by substituting u_t for u_{tt} in equation (3.5.25) and the condition $u(0) = u_0$ for condition (3.5.26) (the case of a first order equation).

Remark 3.5.2. Using the assertions of Theorems 3.5.1 and 3.5.2 applied to certain operators given in the examples below, it is possible to obtain similar results not only for scalar functions $f(t)$, but for any $f(x,t) \in L_2(0,T; L_2(D))$ satisfying the corresponding conditions (see Amirov, 1987).

Examples

1. Let $H = L_2(D)$, where D is a bounded domain in $\mathbb{R}^n$ with boundary $\partial D \in C^m$. The differential expression

$$lu \equiv \sum_{|p|,|q|\leq m} (-1)^{|p|} \frac{\partial^p}{\partial x^p}\Big(a_{pq}\frac{\partial^q u}{\partial x^q}\Big),$$

where $a_{pq} = a_{qp} \in C^m(D)$, naturally generates a symmetric operator l_0 on $C^{2m} \cap C_0^{m-1}(D)$. Here $C_0^{m-1}(D)$ is the set of functions differentiable up to the order of $m-1$ and vanishing on ∂D. We denote the closure of l_0 by L. It is known that the operator L satisfies the conditions imposed on the operator A in Theorems 3.5.1 and 3.5.2 (see Lions and Magenes, 1972).

2. Let $D \subset \mathbb{R}^3$ and $\partial D \in C^2$. The operator

$$lu \equiv -(\lambda+\mu)\operatorname{grad}\operatorname{div} u - \mu\Delta u$$

with boundary conditions

$$\sum_{i=1}^{3} n_i \sigma_{ij} = 0 \quad \text{on } \partial D,$$

where $\sigma_{ij} = \lambda(\operatorname{div} u)\delta_{ij} + \mu\Big(\frac{\partial u_i}{\partial x_j} - \frac{\partial u_j}{\partial x_i}\Big)$ is the stress tensor, satisfies the conditions for the operator A in Theorems 3.5.1 and 3.5.2 (see Taylor, 1981). Here $u(x,t)$ assumes values in $\mathbb{R}^3$, and by n_i we denote the components of the unit normal to ∂D. The constants λ and μ are positive.

Chapter 4.

Inverse problems for second order differential equations

This chapter deals with inverse problems for second order differential equations. In Section 4.1, an inverse problem for the quantum kinetic equation is studied. Applying the Fourier transform with respect to p (momenta) to this problem, we obtain an inverse problem for an ultrahyperbolic equation, which is studied in the case of an ultrahyperbolic equation of more general form in Section 4.2.

It is known that inverse problems for differential equations are closely connected with integral geometry (see Romanov, 1972). Section 4.3 is devoted to a class of inverse problems for differential equations of hyperbolic, elliptic, and parabolic types.

The problem analyzed in Section 4.4 is associated with an inverse problem for a hyperbolic equation with the data containing Dirac's delta function. Such problems are connected with the problems of integral geometry on ellipsoids. Furthermore, they are of substantial interest, for example, from the geophysical point of view. The reason is that the data in applied problems related to the problems of exploration geophysics are often represented by finite functions vanishing outside a relatively small domain. In the study of such problems, the mathematical models for such situations involve generalized functions with supports concentrated in a set of fixed points in space (see Romanov, 1984).

4.1. QUANTUM KINETIC EQUATION

By quantum kinetic equations we mean equations of the form

$$\frac{\partial u}{\partial t} = \frac{i}{(2\pi)^n h} \int_{\mathbb{R}^{2n}} \Big[H\Big(x - \frac{h}{2}q,\, p + \frac{h}{2}s\Big) - H\Big(x + \frac{h}{2}q,\, p - \frac{h}{2}s\Big)\Big] \exp\left\{\mathrm{i}\left[s(x'-x) + q(p'-p)\right]\right\}$$
$$\times\, u(x', p', t)\, \mathrm{d}s\, \mathrm{d}q\, \mathrm{d}x'\, \mathrm{d}p' + stu + f. \tag{4.1.1}$$

Here $u(x,p,t)$ is the quantum distribution function, h is Planck's constant, i is the imaginary unit, $H(x,p)$ is the Hamiltonian function, stu is the collision integral, $f(x,p,t)$ is the function characterizing the sources. By expansion in power series of h, equation (4.1.1) can be reduced to the classical approximate form

$$\frac{\partial u}{\partial t} + \{u, H\} = stu + f, \tag{4.1.2}$$

where $\{u, H\}$ is the Poisson bracket of u and H. In particular, for a special choice of stu, from (4.1.2) we can obtain the Boltzmann equation. From the latter, using the H-theorem, it is possible to derive different transport equations.

If stu is known and $H = \frac{1}{2}|p|^2 + \Phi(x)$, equation (4.1.1) implies

$$\frac{\partial u}{\partial t} + \sum_{j=1}^{n} p_j \frac{\partial u}{\partial x_j} = \frac{i}{(2\pi)^n h} \int_{\mathbb{R}^{2n}} \Big[\Phi\Big(x - \frac{1}{2}hy, t\Big) - \Phi\Big(x + \frac{1}{2}hy, t\Big)\Big]$$
$$\times \exp\left[\mathrm{i}y(p - p')\right] u(x, p', t)\, \mathrm{d}p'\, \mathrm{d}y + f. \tag{4.1.3}$$

Equation (4.1.3) is considered in the domain

$$\Omega = \left\{(x,p,t) \,:\, x_1 > 0, {}'x \in \mathbb{R}^{n-1}, p \in \mathbb{R}^n, t \in \mathbb{R}\right\}.$$

We now formulate the following inverse problem.

Problem 4.1.1. Find the potential $\Phi(x,t)$ in the kinetic equation (4.1.3) and the solution $u(x,p,t)$ of (4.1.3), provided that the following conditions are specified:

$$u(x,p,t)\big|_{x_1\le 0} = u_1(x,p,t), \quad u(x,p,0) = u_0(x,p). \tag{4.1.4}$$

Before we formulate the main result of this section, some preparatory notation and operations will be introduced. Applying the Fourier transform with respect to p, from (4.1.3) we obtain the equation:

$$\frac{\partial \hat{u}}{\partial t} + \mathrm{i}\sum_{j=1}^{n} \frac{\partial^2 \hat{u}}{\partial x_j \partial y_j} - A\hat{u} = \hat{f}, \tag{4.1.5}$$

where $A = \frac{\mathrm{i}}{h}\Big[\Phi\Big(x - \frac{1}{2}hy, t\Big) - \Phi\Big(x + \frac{1}{2}hy, t\Big)\Big]$. From (4.1.4) we find

$$\hat{u}\big|_{x_1<0} = \hat{u}_1, \quad \hat{u}(x,y,0) = \hat{u}_0(x,y). \tag{4.1.6}$$

Making the change of variables

$$x - \frac{1}{2}hy = \xi, \quad x + \frac{1}{2}hy = \eta$$

and introducing the function $w(\xi,\eta,t) = \hat{u}(x,y,t)$ in (4.1.5), we get

$$w_t + \frac{\mathrm{i}}{2}h(\Delta_\eta - \Delta_\xi)w + \mathrm{i}\,[\Phi(\eta) - \Phi(\xi)]\,w = \hat{f}, \tag{4.1.7}$$

and from (4.1.6) it follows that

$$w(\xi,\eta,0) = w_0, \quad w(\xi,\eta,t)\big|_{\xi_1+\eta_1\le 0} = w_1. \tag{4.1.8}$$

We introduce the sets $\mathfrak{M}_u$ and $\mathfrak{M}_\Phi$ as follows. $\mathfrak{M}_u$ is the set of functions $\hat{u}$ such that $\hat{u} \in C^2 \cap H^{(n+2)/2}(\mathbb{R}^{2n+1})$, $\hat{u} = 0$ for $x_1 - \frac{1}{2}hy_1 \ge \xi_0$, and the Fourier transform of $\hat{u}$ with respect to t is finite, with support in $(-R_0, R_0)$, $R_0 > 0$. The set $\mathfrak{M}_\Phi$ consists of finite functions $\Phi \in C(\mathbb{R}^n)$ whose supports lie in the half-space $\xi_1 < \xi_0 - \varepsilon$ $(\varepsilon > 0)$.

Theorem 4.1.1. *Let $\hat{u}_0 \ne 0$ for $x_1 - \frac{1}{2}hy_1 \ge \xi_0$. Then Problem 4.1.1 has at most one solution (u, Φ) such that $u \in \mathfrak{M}_u$ and $\Phi \in \mathfrak{M}_\Phi$.*

Proof. Let (u_1, Φ_1) and (u_2, Φ_2) be two solutions of Problem 4.1.1 such that $(u_i, \Phi_i) \in \mathfrak{M}_u \times \mathfrak{M}_\Phi$, $i = 1, 2$. For $\hat{u} = u_2 - u_1$, $\Phi = \Phi_2 - \Phi_1$, $\hat{u}(x, y, t) = w(\xi, \eta, t)$, from (4.1.1) it follows that

$$w_t + \frac{\mathrm{i}}{2} h(\Delta_\eta - \Delta_\xi) w + \mathrm{i}\,[\Phi(\eta) - \Phi(\xi)]\,\hat{u}_1 + \mathrm{i}\,[\Phi_2(\eta) - \Phi_2(\xi)]\,w = 0, \tag{4.1.9}$$

where Δ_η is the Laplacian with respect to η.

Since $\hat{u} = 0$ for $x_1 - hy_1/2 \geq \xi_0$, by virtue of (4.1.8) we have

$$w(\xi, \xi_0, '\eta, t) = w'_{\eta_1}(\xi, \xi_0, '\eta, t) = 0, \quad w(\xi, \eta, 0) = 0. \tag{4.1.10}$$

Setting $t = 0$ in (4.1.9) and taking into account condition (4.1.10), we obtain

$$\Phi(\eta) - \Phi(\xi) = \frac{1}{\hat{u}_0} \int_{\mathbb{R}} \omega \tilde{w} \, d\omega, \tag{4.1.11}$$

where

$$\tilde{w} = \int_{\mathbb{R}} w(\xi, \eta, t) \exp(\mathrm{i} t\omega) \, dt \tag{4.1.12}$$

Applying the Fourier transform with respect to (ξ, t) to the problem (4.1.9) – (4.1.10) and recalling (4.1.11), we get

$$\omega \hat{w} + \frac{1}{2} h \Delta_\eta \hat{w} + |s|^2 \frac{h}{2} \hat{w}$$
$$- a * \int_{\mathbb{R}} \omega \hat{w}(s, \eta, \omega) \, d\omega + \Phi_2(\eta) \hat{w} - (\Phi_2 * w)_\xi = 0, \tag{4.1.13}$$

where

$$\hat{w}(s, \eta, \omega) = \int_{\mathbb{R}^{n+1}} w(\xi, \eta, t) \exp[\mathrm{i}(t\omega + s\xi)] \, d\xi \, dt,$$

$a(\xi, \eta, \omega) = \hat{u}_1(\xi, \eta, \omega)/\hat{u}(\xi, \eta, 0)$, and $*$ is the convolution with respect to ξ.

It turns out that if $\hat{w}$ is a solution to the problem (4.1.12) – (4.1.13) that belongs to $\mathfrak{M}_u$, then $\hat{w} = 0$. This fact is a special case that follows from the proof of Theorem 4.2.1 (see Remark 4.2.2), since $w(\xi, \eta, t) = \hat{u}(x, y, t)$ and the Fourier transform of $\hat{u}$ with respect to t is finite, its support being contained in $(-R_0, R_0)$.

Therefore, taking $\xi_1 \in (\xi_0 - \varepsilon, \xi_0)$ in equation (4.1.9) and observing that $u_0 \neq 0$, we get $\Phi(\eta) = 0$. $\square$

4.2. ULTRAHYPERBOLIC EQUATION

This section is devoted to inverse problems for ultrahyperbolic equations. An equation of this type appeared before in Section 4.1 in the context of the study of an inverse problem for the quantum kinetic equation. Here we shall consider an ultrahyperbolic equation of more general form.

Let $\Omega = \{(x,y) : x \in D, y \in R^m\}$, where $D \subset R^n$ satisfies the "cone condition" (see Section 2.6). Consider the following equation in Ω:

$$Lu \equiv \Delta_y u - c^2\Delta_x u + \sum_{i=1}^{n} a_i u_{x_i} + \sum_{i=1}^{m} b_i u_{y_i} + a_0 u = f(x,y)g(x). \quad (4.2.1)$$

Here $m,n \geq 1$, $c \in C^1(\overline{D})$, the functions a_j, b_i ($j = 0,1,\dots,n$; $i = 0,1,\dots,m$) are measurable in Ω and have generalized derivatives with respect to y such that for any $x \in D$ and multi-index β ($|\beta| \leq \alpha+1$) we have $|D_y^\beta a_j| \leq M$, $|D_y^\beta b_j| \leq M$, where M is a positive constant,

$$\alpha = \begin{cases} m/2+1 & \text{for even } m, \\ (m+1)/2 & \text{for odd } m. \end{cases}$$

Moreover, for any fixed $x \in D$ we assume that $f \in H^{\alpha+1}(\mathbb{R}^m)$ and $f \in C(\Omega)$ as a function of y.

In the sequel we write $'x = (x_2, x_3, \dots, x_n)$, $x' = (x_1, x_2, \dots, x_{n-1})$.

Problem 4.2.1. Determine the pair (u,g) from equation (4.2.1), provided that for the solution to the Cauchy problem for (4.2.1) with the data

$$u(0,'x,y) = u_{x_1}(0,'x,y) = 0 \quad (4.2.2)$$

the following additional information is given:

$$u(x,0) = u_0(x). \quad (4.2.3)$$

We introduce the set $\mathfrak{M}$ of functions $\varphi(x,y) \in C^2(\overline{\Omega})$ such that $\varphi, \varphi_{x_i} \in H^{\alpha+2}(\mathbb{R}^m)$ as functions of y for any fixed $x \in D$, $i = 1,2,\dots,n$; $\int \xi_j \hat{u}\, d\xi < \infty$, $j = 1,2,\dots,m$, $\int |\xi|^2 \hat{u}\, d\xi < \infty$.

Theorem 4.2.1. *Let $c \in C^1(\overline{D})$, $c > 0$, $c_{x_1} > 0$, and $f(x,0) \neq 0$ on $\overline{D}$. Then Problem 4.2.1 has at most one solution (u,g) such that $u \in \mathfrak{M}$ and $g \in L_1(D)$.*

Proof. Assume that $u_0(x) = 0$ and the pair (u, g) is a solution to Problem 4.2.1 such that $u \in \mathfrak{M}$ and $g \in L_1(D)$. We can put $y = 0$ in (4.2.1), since the corresponding traces exist, as follows from the conditions imposed on the functions a_i, b_i, a_0, f. Then, observing that $u_0 = 0$, from (4.2.1) we get

$$g(x) = \frac{1}{f(x,0)}\Big[\Delta_y u(x,0) + \sum_{j=1}^{m} b_j(x,0)u_{y_j}(x,0)\Big],$$

or, equivalently,

$$g(x) = \frac{1}{f(x,0)}\Big(-\int_{\mathbb{R}^m} |\xi|^2\hat{u}(x,\xi)\,\mathrm{d}\xi + \mathrm{i}\sum_{j=1}^{m} b_j(x,0)\int_{\mathbb{R}^m} \xi_j\hat{u}(x,\xi)\,\mathrm{d}\xi\Big),$$

where $\hat{u}$ is the Fourier transform of $u(x, y)$ with respect to y and i is the imaginary unit.

Dividing both sides of equation (4.2.1) by $c^2(x)$ and applying the Fourier transform with respect to y, we obtain

$$-|\xi|^2c^{-2}\hat{u} - \Delta_x\hat{u} + \sum_{j=1}^{n}\widehat{a_ju_{x_j}} + \sum_{j=1}^{m}\widehat{b_ju_{y_j}} + \widehat{a_0u}$$
$$= \hat{f}(x,\xi)\frac{1}{f(x,0)}\Big(-\int_{\mathbb{R}^m}|\xi|^2\hat{u}\,\mathrm{d}\xi + \mathrm{i}\sum_{j=1}^{m}\int_{\mathbb{R}^m}\xi_j\hat{u}\,\mathrm{d}\xi\Big), \quad (4.2.4)$$

where $\widehat{a_ju_{x_j}}$ is the Fourier transform of $a_ju_{x_j}$ with respect to y. For brevity, a_j and b_j denote a_j/c^2 and b_j/c^2, respectively.

From condition (4.2.2) it follows that

$$\hat{u}(0,'x,\xi) = \hat{u}_{x_1}(0,'x,\xi) = 0. \quad (4.2.5)$$

We now prove that the Cauchy problem for equation (4.2.4) with data (4.2.5) has only trivial solution in $\mathfrak{M}$. To this end, we prove that $u = 0$ and $g = 0$ for all $y \in \mathbb{R}^m$ in some neighbourhood $\Omega(x_0)$ of any point $x_0 \in D$ such that $0 < x_1^0 < \delta$.

The following notation is introduced: $L_0\hat{u} \equiv -(|\xi|^2c^{-2} + \Delta_x)\hat{u}$;

$$\Omega(x_0) = \{x \in \mathbb{R}^n \ :\ x_1 > 0,\ \ 0 < \delta x_1 < \gamma - \frac{1}{2}\sum_{i=2}^{n}(x_i - x_i^0)^2\},$$

where $\gamma < 1$, $\delta > 1$, $x_0 = (x_1^0, \dots, x_n^0)$, and $\gamma + \alpha_0 = \eta < 1$;

$$\psi(x) = \delta x_1 + \frac{1}{2}\sum_{i=2}^{n}(x_i - x_i^0)^2 + \alpha_0, \quad \alpha_0 > 0.$$

Obviously, $x_0 \in \Omega(x_0)$ and $\Omega(x_0) \subset \Omega$ for sufficiently small $\gamma > 0$. Furthermore, $\varkappa = \exp(\lambda\psi^{-\nu})$, $\alpha_0 < \psi(x) < \eta$, where λ and ν are positive parameters satisfying some additional conditions to be specified later.

The proof of the following lemma is trivial.

Lemma 4.2.1. *The following equality holds for any function $u \in \mathfrak{M}$:*

$$\begin{aligned} -(\varphi(L_0\varphi))\varkappa^2 = -c^{-2}|\xi|^2\varphi^2 + |\nabla_x\varphi|^2 \\ - \varphi^2\sum_{i=1}^{n}(2\lambda^2\nu^2\psi_{x_i}^2\psi^{-2\nu-2} + \lambda\nu(\nu+1)\psi^{-\nu-2}\psi_{x_i} \\ - \lambda\nu\psi^{-\nu-1}\psi_{x_ix_i})\varkappa^2 + d_1(\varphi), \end{aligned} \tag{4.2.6}$$

where

$$d_1(\varphi) = -\sum_{i=1}^{n}\big((\varphi\varphi_{x_i} + \psi^{-\nu-1}\lambda\nu\psi_{x_i}\varphi^2)\varkappa^2\big)_{x_i}$$

and φ is the real (or imaginary) component of $\hat{u}$.

Lemma 4.2.2. *Let $c \in C^1(\overline{D})$, $c > 0$, $c_{x_1} > 0$ on $\overline{D}$, $\delta > \delta_0$, where $\delta_0 = \max\{1, \bar{\delta}_0\}$,*

$$\bar{\delta}_0 = \max_{x\in\overline{D}}\Big\{\sum_{i=2}^{n}|c_{x_i}(x_i - x_i^0)| + nC\Big\}\Big/\min_{x\in\overline{D}} C_{x_1},$$

and let positive parameters ν and λ satisfy the following conditions:

$$\nu \geq n + \nu_0, \qquad \nu_0 = \frac{\eta(n+1)}{\delta^2(1-\eta)},$$

$$\lambda \geq \lambda_0 = 2\big[(\nu+2)^2(\delta^2+\gamma^2)^2 + (n-1)^2\big]$$

Then the following estimate is valid (in the notation of Lemma 4.2.1):

$$\begin{aligned} &\psi^{\nu+1}(L_0\varphi)^2\varkappa^2 \\ &\geq 3\lambda^3\nu^4\delta^4\psi^{-2\nu-2}\varphi^2\varkappa^2 + 2\lambda\nu(n+1)|\xi^2|c^{-2}\varphi^2\varkappa^2 \\ &\quad - 2\lambda\nu(n-1)|\nabla_x\varphi|^2\varkappa^2 + d_2(\varphi) \end{aligned} \tag{4.2.7}$$

where $d_2(\varphi) = \sum_{i=1}^{5} d_{2i}(\varphi)$,

$$d_{21}(\varphi) = 2\lambda\nu|\xi|^2 \sum_{i=1}^{n} (\psi_{x_i} c^{-2} \varphi^2)_{x_i},$$

$$d_{22}(\varphi) = 4\lambda\nu \sum_{i,j=1}^{n} \left[\psi_{x_i}(\varphi_{x_i} - \lambda\nu\psi_{x_i}\psi^{-\nu-1}\varphi)(\varphi_{x_j} - \lambda\nu\psi_{x_j}\psi^{-\nu-1}\varphi)\varkappa^2\right]_{x_j}$$

$$- 2\lambda\nu \sum_{i,j=1}^{n} \left[\psi_{x_i}(\varphi_{x_j} - \lambda\nu\psi_{x_j}\psi^{-\nu-1}\varphi)^2\varkappa^2\right]_{x_i}$$

$$+ 2\lambda^2\nu^2(n-1) \sum_{i=1}^{n} (\psi^{-\nu-1}\psi_{x_i}\varphi^2\varkappa^2)_{x_i},$$

$$d_{23}(\varphi) = 2\lambda^3\nu^3 \Big(\psi^{-2\nu-2}\varphi^2|\nabla_x\psi|^2 \sum_{i=1}^{n} \psi_{x_i}\Big)_{x_i},$$

$$d_{24}(\varphi) = -2\lambda^2\nu^2(\nu+1) \sum_{i=1}^{n} (|\nabla_x\psi|^2\psi^{-\nu-2}\psi_{x_i}\varphi^2)_{x_i},$$

$$d_{25}(\varphi) = 2\lambda^2\nu^2(n-1) \sum_{i=1}^{n} (\psi_{x_i}\psi^{-\nu-1}\varphi^2)_{x_i},$$

Proof. The proof of Lemma 4.2.2 is a modification of the proof of Lemma 2 from Lavrent'ev, Romanov, and Shishatskii (1980), p. 82. We introduce the auxiliary function $v(x,\xi) = \varkappa\varphi$. As a result, we have

$$\varphi_{x_i} = \varkappa^{-1}(v_{x_i} + \lambda\nu\psi^{-\nu-1}\psi_{x_i}v),$$

$$\varphi_{x_ix_i} = \varkappa^{-1}(v_{x_ix_i} + 2\lambda\nu\psi_{x_i}\psi^{-\nu-1}v_{x_i} + \lambda\nu\psi_{x_ix_i}\psi^{-\nu-1}v$$

$$- \lambda\nu(\nu+1)\psi_{x_i}^2\psi^{-\nu-2}v + \lambda^2\nu^2\psi_{x_i}^2\psi^{-2\nu-2}v).$$

It is obvious that

$$\psi^{\nu+1}\varkappa^2(c^{-2}|\xi|^2\varphi + \Delta_x\varphi)^2$$

$$= \psi^{\nu+1}(c^{-2}|\xi|^2 v + \Delta v) + \Big(\sum_{i=1}^{n} [\lambda^2\nu^2\psi^{-2\nu-2}\psi_{x_i}^2$$

$$- \lambda\nu(\nu+1)\psi^{-\nu-2}\psi_{x_i}^2 + \lambda\nu\psi^{-\nu-1}\psi_{x_ix_i}]v + 2\lambda\nu\psi^{-\nu-1}\sum_{i=1}^{n}\psi_{x_i}v_{x_i}\Big)^2$$

$$\geq 4\lambda\nu\Big(\sum_{i=1}^{n}\psi_{x_i}v_{x_i}\Big)\Big(c^{-2}|\xi|^2 v + \Delta_x v$$

$$+ \sum_{j=1}^{n}\lambda\nu\psi^{-\nu-1}[\lambda\nu\psi^{-\nu-1}\psi_{x_j}^2 - (\nu+1)\psi^{-1}\psi_{x_j}^2 + \psi_{x_jx_j}]v\Big). \qquad (4.2.8)$$

Each term in this expression will be estimated separately.

The first term:

$$4\lambda\nu c^{-2}|\xi|^2 v\sum_{i=1}^{n}\psi_{x_i}v_{x_i}$$

$$= 2\lambda\nu|\xi|^2\sum_{i=1}^{n}(\psi_{x_i}c^{-2}v^2)_{x_i} + 4\lambda\nu|\xi|^2 c^{-3}v^2\Big(\delta c_{x_1} + \sum_{i=2}^{n}c_{x_i}(x_i - x_i^0)\Big)$$

$$- 2\lambda\nu(n-1)c^{-2}|\xi|^2 v^2 \geq d_{21}(v) + 2\lambda\nu|\xi|^2 c^{-2}v^2 \geq d_{21}(v)$$

$$+ 2\lambda\nu(n+1)|\xi|^2 c^{-2}v^2$$

$$= d_{21}(\varkappa\varphi) + 2\lambda\nu(n+1)|\xi|^2 c^{-2}\varkappa^2\varphi^2, \qquad (4.2.9)$$

because

$$\delta > \max_{x\in\overline{D}}\Big\{\sum_{i=2}^{n}|c_{x_i}(x_i - x_i^0)| + nC\Big\}\Big/\min_{x\in\overline{D}}c_{x_1}.$$

The second term:

$$4\lambda\nu\sum_{i,j=1}^{n}\psi_{x_i}v_{x_i}v_{x_jx_j}$$

$$= 4\lambda\nu\sum_{i,j=1}^{n}(\psi_{x_i}v_{x_i}v_{x_j})_{x_j} - 2\lambda\nu\sum_{i,j=1}^{n}(\psi_{x_i}v_{x_j}^2)_{x_i} - 2\lambda\nu\sum_{i=1}^{n}v_{x_i}^2 - 4\lambda\nu\sum_{i=2}^{n}v_{x_i}^2$$

We now write this equality for φ. Since $v_{x_i} = \varkappa(\varphi_{x_i} - \lambda\nu\psi^{-\nu-1}\psi_{x_i}\varphi)$,

$$4\lambda\nu\sum_{i,j=1}^{n}\psi_{x_i}v_{x_i}v_{x_jx_j} \tag{4.2.10}$$

$$= d_{22}(\varkappa\varphi) - 2\lambda\nu(n-1)|\nabla_x\varphi|^2\varkappa^2 - 2\lambda^3\nu^3(n-1)\psi^{-2\nu-2}|\nabla_x\psi|^2\varphi^2\varkappa^2$$

$$+ 2\lambda^2\nu^2(\nu+1)(n-1)|\nabla_x\psi|^2\psi^{-\nu-2}\varphi^2\varkappa^2 - 2\lambda^2\nu^2(n-1)^2\psi^{-\nu-1}\varphi^2\varkappa^2$$

$$\geq d_{22}(\varkappa\varphi) - 2\lambda\nu(n-1)|\nabla_x\psi|^2\varkappa^2 - 2\lambda^3\nu^3(n-1)\psi^{-2\nu-2}|\nabla_x\psi|^2\varphi^2\varkappa^2.$$

The inequality above was obtained using the conditions $\nu \geq n$, $\delta > 1$.

The third term:

$$4\lambda\nu\sum_{i,j=1}^{n}\psi_{x_i}v_{x_i}\lambda^2\nu^2\psi^{-2\nu-2}\psi_{x_j}^2 v \tag{4.2.11}$$

$$= 2\lambda^3\nu^3\sum_{i,j=1}^{n}(\psi_{x_i}\psi_{x_j}^2\psi^{-2\nu-2}v^2)_{x_i} + 4\lambda^3\nu^3(\nu+1)\psi^{-2\nu-3}|\nabla_x\psi|^4v^2$$

$$- 2(n-1)\lambda^3\nu^3\psi^{-2\nu-2}v^2|\nabla_x\psi|^2 - 4\lambda^3\nu^3\psi^{-2\nu-2}v^2|\nabla'_x\psi|^2$$

$$\geq d_{23}(v) + 4\lambda^3\nu^4\psi^{-2\nu-2}|\nabla_x\psi|^2v^2\left(\psi^{-1} - \frac{n+1}{2\nu}|\nabla_x\psi|^2\right)$$

$$\geq d_{23}(v) + 4\lambda^3\nu^4\delta^4\psi^{-2\nu-2}v^2$$

$$= d_{23}(\varkappa\varphi) + 4\lambda^3\nu^4\delta^4\psi^{-2\nu-2}\varphi^2\varkappa^2,$$

because

The fourth term:

$$-4\lambda\nu \sum_{i,j=1}^{n} \psi_{x_i} v_{x_i} \lambda\nu(\nu+1)\psi^{\nu-2}\psi_{x_j}^2 v \tag{4.2.12}$$

$$= -2\lambda^2\nu^2(\nu+1)\sum_{i,j=1}^{n}(\psi_{x_i}\psi_{x_j}^2\psi^{-\nu-2}v^2)_{x_i}$$

$$+ 2\lambda^2\nu^2(\nu+1)(n-1)|\nabla_x\psi|^2\psi^{-\nu-2}v^2$$

$$+ 4\lambda^2\nu^2(\nu+1)|\nabla'_x\psi|^2\psi^{-\nu-2}v^2$$

$$- 2\lambda^2\nu^2(\nu+1)(\nu+2)\psi^{-\nu-3}|\nabla_x\psi|^4 v^2$$

$$\geq d_{24}(v) - 2\lambda^2\nu^2\psi^{-\nu-3}(\nu+1)(\nu+2)|\nabla_x\psi|^4 v^2$$

$$= d_{24}(\varkappa\varphi) - 2\lambda^2\nu^2(\nu+1)(\nu+2)\psi^{-\nu-3}|\nabla_x\psi|^4\varphi^2\varkappa^2.$$

The fifth term:

$$4\lambda\nu \sum_{i,j=1}^{n} \psi_{x_i} v_{x_i} \lambda\nu\psi^{-\nu-1}\psi_{x_j x_j} v$$

$$= 2\lambda^2\nu^2(n-1)\sum_{i=1}^{n}(\psi_{x_i}\psi^{-\nu-1}v^2)_{x_i} - 2\lambda^2\nu^2(n-1)^2\psi^{-\nu-1}v^2$$

$$+ 2\lambda^2\nu^2(n-1)(\nu+1)\psi^{-\nu-2}|\nabla'_x\psi|^2 v^2$$

$$\geq d_{25}(v) - 2\lambda^2\nu^2(n-1)^2\psi^{-\nu-1}v^2$$

$$= d_{25}(\varkappa\varphi) - 2\lambda^2\nu^2(n-1)^2\psi^{-\nu-1}\varkappa^2\varphi^2. \tag{4.2.13}$$

Taking into account inequalities (4.2.8) – (4.2.13), we have

$$\psi^{\nu+1}\varkappa^2(c^{-2}|\xi|^2\varphi + \Delta_x\varphi)^2$$

$$\geq d_2(\varphi) - 2\lambda\nu(n-1)|\nabla_x\varphi|^2\varkappa^2 + 2\lambda\nu(n+1)|\xi|^2c^{-2}\varkappa^2\varphi^2$$

$$+ \lambda^2\nu^2\varphi^2\varkappa^2[4\lambda\nu^2\delta^4\psi^{-\nu-1} - 2(\nu+1)(\nu+2)\psi^{-1}|\nabla_x\psi|^4$$

$$- 2(n-1)^2]\psi^{-\nu-1}. \tag{4.2.14}$$

Since for $\lambda \geq \lambda_0$ we have

$$\lambda\nu^2\delta^4\psi^{-\nu-1} - 2(\nu+1)(\nu+2)\psi^{-1}|\Delta_x\psi|^4 - 2(n-1)^2 \geq 0,$$

inequality (4.2.14) implies (4.2.7), which proves the lemma. □

Lemma 4.2.3. *Under the assumptions of Lemma 4.2.2, suppose that* $\nu \geq \nu_1 = \max\{\nu_0 + n, 4n + 8n(\delta^2 + \eta^2) + 1\}$. *Then*

$$2n\lambda\nu\varphi(L_0\varphi)\varkappa^2 + \psi^{\nu+1}(L_0\varphi)^2\varkappa^2$$
$$\geq 2\lambda^3\nu^3\psi^{-2\nu-2}\varphi^2\varkappa^2 + 2\lambda\nu c^{-2}|\xi|^2\varphi^2\varkappa^2$$
$$+ 2\lambda\nu|\nabla_x\varphi|^2\varkappa^2 + 2n\lambda\nu d_1(\varphi) + d_2(\varphi). \qquad (4.2.15)$$

Proof. Multiply (4.2.6) by $2n\lambda\nu$ and add the obtained expression to (4.2.7). We have

$$2n\lambda\nu\varphi(L_0\varphi)\varkappa^2 + \psi^{\nu+1}(L_0\varphi)^2\varkappa^2$$
$$\geq 2\lambda\nu c^{-2}|\xi|^2\varphi^2\varkappa^2 + 2\lambda\nu|\nabla_x\varphi|^2\varkappa^2 + \big[3\lambda^3\nu^4\delta^4\psi^{-2\nu-2}$$
$$- 4n\lambda^3\nu^3\psi^{-2\nu-2}|\nabla_x\psi|^2 - 2n\lambda^2\nu^2(\nu+1)\psi^{-\nu-2}\sum_{i=1}^{n}\psi_{x_i}$$
$$+ 2n\lambda^2\nu^2\psi^{-\nu-1}(n-1)\big]\varphi^2\varkappa^2 + 2n\lambda\nu d_1(\varphi) + d_2(\varphi).$$

Since $\nu \geq \nu_1$, the foregoing inequality immediately implies (4.2.15), which proves Lemma 4.2.3. □

Proceeding with the proof of Theorem 4.2.1, we introduce the following notation for the problem (4.2.4) – (4.2.5): $\hat{u} = u_1 + iu_2$, $\widehat{a_j u_{x_j}} = a_{1j} + ia_{2j}$, $\widehat{b_j u_{y_j}} = b_{1j} + ib_{2j}$, $\widehat{a_0 u} = a_{10} + ia_{20}$, $\hat{f} = f_1 + if_2$, where a_{sj}, b_{sr}, f_s, u_s ($s = 1,2$; $j = 0,1,\dots,n$; $r = 1,2,\dots,m$) are real-valued functions of x and ξ. With this notation, the problem (4.2.4) – (4.2.5) becomes the Cauchy problem for the system

$$-|\xi|^2c^{-2}u_s - \Delta_x u_s = m_s, \qquad s = 1,2, \qquad (4.2.16)$$

with the data

$$u_s(0,'x,\xi) = u'_{sx_1}(0,'x,\xi) = 0, \qquad (4.2.17)$$

where

$$m_1 = -\sum_{j=0}^{n} a_{1j} - \sum_{j=1}^{m} b_{1j} + \frac{1}{f(x,0)}\Big[f_2 \int_{\mathbb{R}^m} |\xi|^2 u_2 \, \mathrm{d}\xi - f_1 \int_{\mathbb{R}^m} |\xi|^2 u_1 \, \mathrm{d}\xi$$

$$- \sum_{j=1}^{m} b_j(x,0)\Big(f_1 \int_{\mathbb{R}^m} \xi_j u_2 \, \mathrm{d}\xi + f_2 \int_{\mathbb{R}^m} \xi_j u_1 \, \mathrm{d}\xi\Big)\Big],$$

$$m_2 = -\sum_{j=0}^{n} a_{2j} - \sum_{j=1}^{m} b_{2j} + \frac{1}{f(x,0)}\Big[-f_1 \int_{\mathbb{R}^m} |\xi|^2 u_2 \, \mathrm{d}\xi - f_2 \int_{\mathbb{R}^m} |\xi|^2 u_1 \, \mathrm{d}\xi$$

$$+ \sum_{j=1}^{m} b_j(x,0)\Big(f_1 \int_{\mathbb{R}^m} \xi_j u_1 \, \mathrm{d}\xi - f_2 \int_{\mathbb{R}^m} \xi_j u_2 \, \mathrm{d}\xi\Big)\Big],$$

For notation simplicity, we introduce $|\hat{\xi}| = 1 + |\xi|^2$, $\alpha = m/2 + 1$ for even m and $\alpha = (m+1)/2$ for odd m.

The following estimates will be needed in the sequel:

$$\Big(f_s \int_{\mathbb{R}^m} |\xi|^4 \, u_r \, \mathrm{d}\xi\Big)^2 = f_s^2 \Big(\int_{\mathbb{R}^m} |\xi|^2 \, |\hat{\xi}|^{\alpha/2} \, |\hat{\xi}|^{-\alpha/2} u_r \, \mathrm{d}\xi\Big)^2$$
$$\leq f_s^2 c(\alpha) \int_{\mathbb{R}^m} |\xi|^4 \, |\hat{\xi}|^{\alpha} u_r^2 \, \mathrm{d}\xi, \tag{4.2.18}$$

$$\Big(b_j(x,0) f_r \int_{\mathbb{R}^m} \xi_j u_s \, \mathrm{d}\xi\Big)^2 \leq c(\alpha) b_j^2(x,0) f_r^2 \int_{\mathbb{R}^m} |\hat{\xi}|^{\alpha} \, |\xi|^2 u_s^2 \, \mathrm{d}\xi,$$

where $s, r = 1, 2$ and $c(\alpha) = \displaystyle\int_{\mathbb{R}^m} |\hat{\xi}|^{-\alpha} \, \mathrm{d}\xi$.

By virtue of (4.2.18) and the inequality

$$\Big(\sum_{i=1}^{n} a_i\Big)^2 \leq n \sum_{i=1}^{n} a_i^2$$

we have

$$m_1^2 + m_2^2 \leq 3(n-1)\sum_{j=0}^{n}\Big(\sum_{r=1}^{2} a_{rj}^2\Big) + 3m\sum_{j=1}^{m}\Big(\sum_{r=1}^{2} b_{rj}^2\Big)$$
$$+ f_0^2 \sum_{r,s=1}^{2} f_r^2 \int_{\mathbb{R}^m} |\xi|^4\, |\hat{\xi}|^\alpha u_s^2\, \mathrm{d}\xi$$
$$+ 2mf_0^2\, |b|^2\Big(\sum_{r,s=1}^{2} f_r^2 \int_{\mathbb{R}^m} |\xi|^4\, |\hat{\xi}|^\alpha u_s^2\, \mathrm{d}\xi\Big), \qquad (4.2.19)$$

where

$$|b|^2 = \sum_{j=1}^{m} b_j^2(x,0), \qquad f_0^2 = \frac{9c(\alpha_0)}{f^2(x,0)}.$$

By Lemma 4.2.3, for $s = 1,2$ we have

$$(m_s^2 + \lambda^2\nu^2 n^2 u_s^2)\varkappa^2 + \psi^{\nu+1}\varkappa^2 m_s^2$$
$$\geq 2\lambda\nu n\varkappa^2 u_s L_0 u_s + \psi^{\nu+1}\varkappa^2 (L_0 u_s)^2$$
$$\geq \lambda^3\nu^3\psi^{-2\nu-2}u_s^2\varkappa^2 + 2\lambda\nu c^{-2}|\xi|^2 u_s^2\varkappa^2 + 2\lambda\nu|\nabla_x u_s|^2\varkappa^2$$
$$+ 2n\lambda\nu d_1(u_s) + d_2(u_s). \qquad (4.2.20)$$

By virtue of inequality (4.2.19) and the equalities $u_1^2 + u_2^2 = |\hat{u}|^2$, $a_{1j}^2 + a_{2j}^2 = |\widehat{a_j u_{x_j}}|^2$, $b_{1j}^2 + a_{2j}^2 = |\widehat{b_j u_{y_j}}|^2$, $a_{10}^2 + a_{20}^2 = |\widehat{a_0 u}|^2$, $f_1^2 + f_2^2 = |\hat{f}|^2$, summing inequalities (4.2.20) with respect to $s = 1,2$, we obtain

$$\Big[3(n+1)|\widehat{a_0 u}|^2 + 3(n+1)\sum_{j=1}^{n} |\widehat{a_j u_{x_j}}|^2$$
$$+ 3m\sum_{j=1}^{m} |\widehat{b_j u_{y_j}}|^2 + f_0|\hat{f}|^2 \int_{\mathbb{R}^m} |\xi|^4\, |\hat{\xi}|^\alpha |\hat{u}|^2\, \mathrm{d}\xi$$
$$+ 2mf_0^2|b|^2\, |\hat{f}|^2 \int_{\mathbb{R}^m} |\xi|^2\, |\hat{\xi}|^\alpha |\hat{u}|^2\, \mathrm{d}\xi + \lambda^2\nu^2 n^2 |\hat{u}|^2\Big]\varkappa^2 \geq$$
$$\geq 2\lambda^3\nu^3\psi^{-2\nu-2}|\hat{u}|^2\varkappa^2 + 2\lambda\nu c^{-2}|\xi|^2|\hat{u}|^2\varkappa^2$$
$$+ 2\lambda\nu|\nabla_x \hat{u}|^2\varkappa^2 + \sum_{s=1}^{2}(2n\lambda\nu d_1(u_s) + d_2(u_s)).$$

Multiplying this inequality by $|\hat{\xi}|^{\alpha+1}$ and integrating the resulting relation over $\mathbb{R}^m$, we get

$$3\varkappa^2 \int_{\mathbb{R}^m} |\hat{\xi}|^{\alpha+1}\Big[(n+1)\Big(|\widehat{a_0 u}|^2 + \sum_{j=1}^{n} |\widehat{a_j u_{x_j}}|^2\Big) + m \sum_{j=1}^{m} |\widehat{b_j u_{y_j}}|^2\Big]\, d\xi$$

$$+ f_0^2 \int_{\mathbb{R}^m} |\hat{\xi}|^{\alpha+1} |\hat{f}|^2\, d\xi \varkappa^2 \int_{\mathbb{R}^m} |\hat{\xi}|^{\alpha} |\xi|^2 (|\xi|^2 + 2m|b|^2) |\hat{u}|^2\, d\xi$$

$$+ \varkappa^2 \lambda^2 \nu^2 n^2 \int_{\mathbb{R}^m} |\hat{\xi}|^{\alpha+1} |\hat{u}|^2\, d\xi - \sum_{i=1}^{2} \int_{\mathbb{R}^m} |\hat{\xi}|^{\alpha+1} (2n\lambda\nu d_1(u_s) + d_2(u_s))\, d\xi$$

$$\geq \int_{\mathbb{R}^m} |\hat{\xi}|^{\alpha+1} \varkappa^2 \Big(|\hat{u}|^2 (2\lambda^3\nu^3\psi^{-2\nu-2} + 2\lambda\nu c^{-2}|\xi|^2) + 2\lambda\nu |\nabla_x \hat{u}|^2\Big)\, d\xi$$

$$= \int_{\mathbb{R}^m} |\hat{\xi}|^{\alpha+1} \varkappa^2 \Big(|\hat{u}|^2 (2\lambda^3\nu^3\psi^{-2\nu-2} + 2\lambda\nu c^{-2}|\hat{\xi}|^2 - 2\lambda\nu c^{-2}) + 2\lambda\nu |\nabla_x \hat{u}|^2\Big)\, d\xi. \tag{4.2.21}$$

The following estimates can be established on the basis of the assumptions of the theorem:

$$\int_{\mathbb{R}^m} |\hat{\xi}|^{\alpha+1} |\widehat{a_j u_{x_j}}|^2\, d\xi \leq C_m \int_{\mathbb{R}^m} \sum_{|\beta|\leq\alpha+1} (D_y^\beta (a_j u_{x_j}))^2\, dy$$

$$\leq C_m M \int_{\mathbb{R}^m} \sum_{|\beta|\leq\alpha+1} (D_y^\beta u_{x_j})^2\, dy \leq C'_m M \int_{\mathbb{R}^m} |\hat{\xi}|^{\alpha+1} |\hat{u}_{x_j}|^2\, d\xi,$$

$$\int_{\mathbb{R}^m} \sum_{j=1}^{m} |\hat{\xi}|^{\alpha+1} |\widehat{b_j u_{y_j}}|^2\, d\xi \leq M \int_{\mathbb{R}^m} |\xi|^{\alpha+1} |\xi|^2 |\hat{u}|^2\, d\xi \tag{4.2.22}$$

Now we note that $\lambda \geq \lambda_0 = \max\{\lambda_1, \lambda_2, \lambda_3\}$, where

$$\lambda_1 = c^2(\bar{f}_0^2 + mM), \quad \lambda_2 = 3n(n+1)C'_m M,$$

$$\lambda_3 = 3(n+1)C'_m M + 2c^{-2} + n^2 + 2m|\bar{b}|^2 + 1,$$

$$\bar{f}_0^2 = \max_{x\in\overline{D}} f_0^2, \quad |\bar{b}|^2 = \max_{x\in\overline{D}} |b|^2.$$

Hence, (4.2.21) – (4.2.22) imply

$$\lambda^3\nu^3 \int_{\mathbb{R}^m} |\hat{u}|^2 \varkappa^2\, d\xi \leq \operatorname{div}(\hat{u}),$$

where

$$\operatorname{div}(\hat{u}) = \sum_{s=1}^{2} \int_{\mathbb{R}^m} |\hat{\xi}|^{\alpha+1} (2n\lambda\nu d_1(u_s) + d_2(u_s))\, d\xi.$$

Thus, on $\Omega(x_0)$ we have

$$\int_{\mathbb{R}^m} |\hat{u}|^2\, d\xi \leq \int_{\mathbb{R}^m} |\hat{u}|^2 \varkappa^2\, d\xi \leq \frac{1}{\lambda^3\nu^3} \operatorname{div}(\hat{u}).$$

Integrating the last inequality over $\Omega(x_0)$ and passing to the limit as $\lambda \to \infty$, we have

$$\int_{\Omega(x_0)} \int_{\mathbb{R}^m} |\hat{u}|^2\, dx\, d\xi \leq 0,$$

which means that $u \equiv 0$ in $\Omega(x_0)$ for $y \in \mathbb{R}^m$.

Varying the point $x_0 = (0, x_2^0, \dots, x_n^0)$ of the surface $x_1 = 0$ we establish that $u = 0$ on $\Omega_\delta = \{(x, y) : x \in D, y \in \mathbb{R}^m, 0 < x_1 < \delta\}$, and from equation (4.2.1) we conclude that $g(x) = 0$ on $D_\delta = \{x : x \in D, 0 < x_1 < \delta\}$. With the previous reasoning repeated for the domain Ω_δ, we establish that $u = 0$ in $\Omega_{2\delta}$ and $g = 0$ on $D_{2\delta}$, and so on. This completes the proof of Theorem 4.2.1. □

Consider the following coefficient inverse problem for equation (4.2.1).

Problem 4.2.2. Determine the function $u(x, y)$ and the coefficient $a_{\alpha_0}(x)$ from equation (4.2.1), provided that all other coefficients and the right-hand side of equation (4.2.1) are given and conditions (4.2.2), (4.2.3) are specified.

The question of solution uniqueness for Problem 4.2.2 is reduced to that for Problem 4.2.1 (see Lavrent'ev, Romanov, and Shishatskii, 1980). Thus, Theorem 4.2.1 implies

Theorem 4.2.2. *Let $c \in C^1(D)$, $c > 0$, $c_{x_1} > 0$, and $\partial u(x, 0)/\partial x_{\alpha_0} \neq 0$ on $\overline{D}$. Then Problem 4.2.2 has at most one solution $(u, a_{\alpha_0}(x))$ such that $u \in \mathfrak{M}$ and $a_{\alpha_0} \in L_1(D)$.*

Remark 4.2.1. The conditions that are specified in the hypotheses of Theorems 4.2.1 and 4.2.2 are essential: $c > 0$ implies ultrahyperbolicity (or hyperbolicity) of equation (4.2.1), and the condition c_{x_1} makes (4.2.2) informative, since for $m = 1$ it implies the absence of waveguides in a medium with $x_1 > 0$ (see Romanov, 1984).

Remark 4.2.2. It is clear from the proof of Theorem 4.2.1 that its assertion is valid also for Problem 4.2.1 if $c = \text{const}$ and $a_j = b_j = 0$ in (4.2.1) for $i = 1, \dots, n$, $j = 1, \dots, m$.

4.3. ON A CLASS OF MULTIDIMENSIONAL INVERSE PROBLEMS

In this section we analyze inverse problems for hyperbolic equations. These problems are known to be closely connected with problems of integral geometry (see Romanov, 1972). Inverse problems for parabolic equations are also considered here. These are the problems to which the corresponding inverse problems for hyperbolic and elliptic equations are reduced.

1. In the first part of this section, we consider an inverse problem for a hyperbolic equation that is associated with Problem 4.2.1 in the case $m = 1$. Throughout this section, by Problem 4.2.1 and equation (4.2.1) we mean their special case for $m = 1$.

For Problem 4.2.1, Theorem 4.2.1 was proved and the physical meaning of the condition $c_{x_1} > 0$ was explained in Remark 4.2.1. We now turn to considering this problem in the case of the domain $y > 0$. In the one-dimensional case ($n = 1$) and the case where the coefficients of equation (4.2.1) are analytic with respect to x_1, in order to provide the uniqueness of solution of Problem 4.2.1 in the domain $y > 0$ it is necessary to add the condition $u_y(x, 0)$, as was shown in Romanov (1984). Thus, we have the following problem.

Problem 4.3.1. Find a solution to Problem 4.2.1 in the domain $y > 0$, provided that $u_y(x, 0) = 0$.

Theorem 4.3.1. *Suppose that the assumptions of Theorem 4.2.1 hold in the domain $y > 0$ and for any function in equation (4.2.1) there exists an even continuation of the same class in the domain $y < 0$. Then Problem 4.3.1 has at most one solution (u, g) such that $u \in \mathfrak{M}_1$, $g \in L_2(D)$, where the definition of $\mathfrak{M}_1$ is similar to the definition of the set $\mathfrak{M}$ from Section 4.2, with the only difference that $\varphi \in H^3(\mathbb{R}_+)$ for any $x \in D$, where $\mathbb{R}_+$ is the positive half of the real axis.*

After taking even continuations of equation (4.2.1) and condition (4.2.2) in the domain $y < 0$ providing $u_y(x, 0) = 0$, the proof of Theorem 4.3.1 is reduced to the proof of Theorem 4.2.1.

Problem 4.3.1 is associated with important applications in acoustics and with the problem of integral geometry (see Romanov, 1972).

Before considering other inverse problems for hyperbolic equations, we analyze the related analogous problems for parabolic equations.

Suppose that Ω satisfies the "cone condition" (see Section 2.6),

$$L = \sum_{i,j=1}^{n} a_{ij}(x,t)\frac{\partial^2}{\partial x_i \partial x_j} + \sum_{i=1}^{n} a_i(x,t)\frac{\partial}{\partial x_i} + a_0(x,t),$$

$$a_{ij} \in C^2(\overline{\Omega}), \quad a_l \in C^1(\Omega), \qquad i,j = 1,\dots,n, \; l = 0,1,\dots,n,$$

the given operator L is uniformly elliptic on Ω, $f(x,t) \in C^1(\Omega)$ is a given function, $\delta > 0$ is as small as desired.

2. The second part begins with the analysis of the following problem:

Problem 4.3.2. Determine the pair of functions (u, φ) in Ω from

$$u_t - Lu = f(x,t)\varphi(x',t), \tag{4.3.1}$$

$$u(0,'x,t) = u_{x_1}(0,'x,t) = 0, \quad x_n \in (-\delta,\delta), \tag{4.3.2}$$

$$u(x',0,t) = g(x't). \tag{4.3.3}$$

Theorem 4.3.2. *Assume that the given functions a_{ij}, a_l, f satisfy the smoothness conditions specified above. Furthermore, suppose $f(x',0,t) \neq 0$ and the functions a_{ij} are independent of x_n.*

Then Problem 4.3.2 has at most one solution (u,φ) such that $u \in C^2(\overline{\Omega})$ and $\varphi \in C(\overline{\Omega})$.

Proof. It suffices to show that the function $\varphi(x',t)$ is uniquely determined from relations (4.3.1) – (4.3.3) in some neighbourhood of an arbitrary point $(x^{0\prime},t^0) \in \Omega^0$, where Ω^0 is the projection of Ω on the surface $x_n = 0$.

We set

$$\Omega_\delta = \Big\{(x,t) \,:\, \delta(1-\varepsilon_0 x_1) - \frac{1}{2}\sum_{i=2}^{n-1}(x_i - x_i^0)^2 - \frac{1}{2}x_n^2 - \frac{1}{2}(t-t_0)^2 - \varepsilon_1 > 0\Big\}.$$

Here $0 < \varepsilon_0 < 1/x_1^0$ for $x_1^0 \geq 1$ and $\varepsilon_0 = 1$ for $x_1^0 < 1$; $0 < \varepsilon_1 < \frac{\delta}{2}(1 - \varepsilon_0 x_1^0)$. Evidently, we have $(x^0,t^0) \in \Omega_\delta$.

Let

$$\psi(x,t) = \delta\varepsilon_0 x_1 + \frac{1}{2}\sum_{i=2}^{n-1}(x_i - x_i^0)^2 - \frac{1}{2}x_n^2 - \frac{1}{2}(t-t_0)^2 + \alpha,$$

where $\delta + \alpha < 1$, $\alpha > 0$.

Since $f(x', 0, t) \neq 0$, there exists a number $0 < \delta < 1$ such that $f \neq 0$ also on Ω_δ. We shall assume that the number δ introduced before satisfies these conditions.

Consider the homogeneous version of Problem 4.3.2 on Ω_δ ($g(x', t) = 0$) and introduce the new unknown function $u = wf$ in Ω_δ. Dividing by f in (4.3.1), we find that $w(x,t)$ satisfies the equation

$$w_t - L'w = \varphi(x', t), \tag{4.3.1$'$}$$

and the conditions

$$w(0, {}'x, t) = w'_{x_1}(0, {}'x, t) = 0, \tag{4.3.2$'$}$$

$$w(0, x', t) = 0 \tag{4.3.3$'$}$$

on Ω_δ, where

$$L' = \sum_{i,j=1}^{n} a_{ij}(x', t)\frac{\partial^2}{\partial x_i \partial x_j} + \sum_{i=1}^{n} a'_i \frac{\partial}{\partial x_i} + a'_0,$$

a'_l ($l = 0, 1, \ldots, n$) are expressed in terms of the coefficients of the operator L and the function f; the principal parts of L' and L coincide.

Differentiating equality (4.3.1$'$) with respect to x_n, putting $\omega = w_{x_n}$, and using (4.3.3$'$), we get

$$\omega_t - L'\omega + \operatorname{sign} x_n \sum_{i=1}^{n} a'_{ix_n} \frac{\partial}{\partial x_i}\int_0^{x_n} \omega(x', \tau, t)\,\mathrm{d}\tau$$
$$+ a'_{0x_n} \operatorname{sign} x_n \int_0^{x_n} \omega(x', \tau, t)\,\mathrm{d}\tau = 0. \tag{4.3.4}$$

From (4.3.2$'$) we obtain

$$\omega(0, {}'x, t) = \omega'_{x_1}(0, {}'x, t) = 0. \tag{4.3.5}$$

We now prove that if $\omega(x,t)$ satisfies conditions (4.3.4) and (4.3.5), then $\omega = 0$ in Ω_δ. To this end, we first consider the Cauchy problem for the differential inequality

$$|\omega_t - L_0\omega| \leq C(|\omega| + |\nabla_x \omega|), \tag{4.3.6}$$

with the conditions

$$\omega(0,'x,t) = \omega'_{x_1}(0,'x,t) = 0, \tag{4.3.7}$$

where

$$L_0 = \sum_{i,j=1}^{n} a_{ij} \frac{\partial^2}{\partial x_i \partial x_j}.$$

Lemma 4.3.1. *There exist an absolute constant λ_0 and a constant ν_0 (depending on δ and the coefficients a_{ij}) such that for $\lambda > \lambda_0$ and $\nu > \nu_0$ and for any $\omega \in C^2(\overline{\Omega})$*

$$c\lambda\nu\varphi^2(\omega_t - L_0\omega)\omega + C\psi^{\nu+2}\varphi^2(\omega_t - L_0\omega)^2 + \widetilde{U}_t + \operatorname{div} U$$

$$\geq \lambda\nu\varphi^2|\nabla_x\omega|^2 + \lambda^3\nu^4\psi^{-2\nu-2}\varphi^2\omega^2$$

where

$$|\widetilde{U}| \leq C\varphi^2(|\nabla_x\omega|^2 + \lambda^2\nu^2\nu_0\psi^{-\nu}\omega^2),$$

$$|U| \leq C\varphi^2(\omega_t^2 + \lambda\nu\nu_0|\nabla_x\omega|^2 + \lambda^3\nu^3\nu_0\psi^{-2\nu-2}\omega^2),$$

$\varphi = \exp(\lambda\psi^{-\nu})$, *$C > 0$ and C is independent of λ, ν.*

Lemma 4.3.1 is proved similarly to Lemma 3 from Lavrent'ev, Romanov, and Shishatskii (1980), p. 93. From Lemma 4.3.1 it follows that

$$\omega = 0 \quad \text{on} \quad \Omega_\delta \tag{4.3.8}$$

Since the free parameters λ and ν in the expression for $\psi(x,t)$ are assumed to be sufficiently large, we note that $\Omega_\delta^{(\nu)} \supset \Omega_\delta^{(1)}$ for all $\nu \geq 1$, where $\Omega_\delta^{(1)} = \Omega_\delta$ and $\Omega_\delta^{(\nu)} = \{(x,t) : \psi^{-\nu} - (\delta+\alpha)^{-\nu} - \varepsilon_1 > 0\}$ for $\nu \geq 2$.

Indeed, let $(x,t) \in \Omega_\delta$. Then $\psi^{-\nu}(x,t) > (\delta+\alpha-\varepsilon_1)^{-\nu}$ and, consequently,

$$\psi^{-\nu}(x,t) - (\delta+\alpha)^{-\nu} - \varepsilon_1$$

$$\geq \frac{(\delta+\alpha)^\nu - (\delta+\alpha-\varepsilon_1)^\nu}{(\delta+\alpha-\varepsilon_1)^\nu(\delta+\alpha)^\nu} - \varepsilon_1 > \frac{\nu\varepsilon_1(\delta+\alpha-\varepsilon_1)^{\nu-1}}{(\delta+\alpha)^\nu(\delta+\alpha-\varepsilon_1)^\nu} - \varepsilon_1$$

$$= \frac{\nu\varepsilon_1}{(\delta+\alpha)^\nu(\delta+\alpha-\varepsilon_1)^\nu} - \varepsilon_1 > (\nu-1)\varepsilon_1 \geq 0,$$

which means that $(x,t) \in \Omega_\delta^{(\nu)}$.

We now prove that the problem (4.3.4) – (4.3.5) is "subordinate" to the problem (4.3.6) – (4.3.7), i.e., the solution to the problem (4.3.4) – (4.3.5) satisfies conditions (4.3.6) – (4.3.7). For this purpose, it suffices to show that

$$\int_{\Omega_\delta} \varphi^2 (A\omega)^2 \, d\Omega_\delta \le \delta^2 \int_{\Omega_\delta} \varphi^2 \omega^2 \, d\Omega_\delta, \tag{4.3.9}$$

where

$$A\omega = \operatorname{sign} x_n \int_0^{x_n} \omega(x', \tau, t) \, d\tau.$$

On Ω_δ we have

$$(A\omega)^2 \le \begin{cases} \displaystyle\int_0^{x_n} \omega^2(x', \tau, t) \, d\tau, & x_n > 0, \\ \displaystyle\int_{x_n}^0 \omega^2(x', \tau, t) \, d\tau, & x_n < 0, \end{cases}$$

and therefore

$$\begin{aligned} &\int_{\Omega_\delta} \varphi^2 (A\omega)^2 \, d\Omega_\delta \\ &\quad = \int_K dx' \, dt \left[\int_0^{\sigma(x',t)} \varphi^2 (A\omega)^2 \, dx_n + \int_{-\sigma(x',t)}^0 \varphi^2 (A\omega)^2 \, dx_n \right] \\ &\quad \le \int_K dx' \, dt \left[\delta \int_0^{\sigma(x',t)} \left(\int_0^{x_n} \varphi^2(x,t) \omega^2(x', \tau, t) \, d\tau \right) dx_n \right. \\ &\qquad \left. + \, \delta \int_{-\sigma(x',t)}^0 dx_n \left(\int_0^{x_n} \varphi^2(x,t) \omega^2(x', \tau, t) \, d\tau \right) \right], \end{aligned} \tag{4.3.10}$$

where K is the section of the domain Ω_δ produced by the plane $x_n = 0$ and $\delta(x', t)$ is the distance between the boundary point (x', x_n, t) of Ω_δ and the point $(x', 0, t) \in K$.

Since φ is monotonic with respect to x_n, for $a = \sigma(x', t)$ we have

$$\begin{aligned} \int_0^a dx_n \left(\int_0^{x_n} \varphi^2(x,t) \omega^2(x', \tau, t) \, d\tau \right) &= \int_0^a d\tau \left(\omega^2(x', \tau, t) \int_\tau^a \varphi^2(x,t) \, dx_n \right) \\ &\le \int_0^a \omega^2(x', \tau, t) \varphi^2(x', \tau, t)(a - \tau) \, d\tau \\ &\le \delta \int_0^a \varphi^2(x', \tau, t) \omega^2(x', \tau, t) \, d\tau. \end{aligned} \tag{4.3.11}$$

Similar reasoning can be used to prove that

$$\int_{-a}^{0} \mathrm{d}x_n \Big(\int_0^{x_n} \varphi^2(x,t)\omega^2(x',\tau,t)\,\mathrm{d}\tau \Big) \le \int_{-a}^{0} \varphi^2(x',\tau,t)\omega^2(x',\tau,t)\,\mathrm{d}\tau. \tag{4.3.12}$$

By virtue of inequalities (4.3.11) and (4.3.12), from (4.3.1) we obtain (4.3.9).

Thus, the problem (4.3.4) – (4.3.5) is "subordinate" to (4.3.6) – (4.3.7). Consequently, (4.3.8) implies $\varphi = w = 0$ in Ω_δ.

The proof of Theorem 4.3.2 is complete. □

Problem 4.3.3. Find a pair of functions (u, φ) satisfying the following relations in Ω:

$$u_t - Lu = f(x,t)\varphi(x) \tag{4.3.13}$$

$$u(x,0) = g_1(x), \tag{4.3.14}$$

$$u(0,'x,t) = 0, \tag{4.3.15}$$

$$u_{x_1}(0,'x,t) = 0, \qquad t \in (-\delta, \delta). \tag{4.3.16}$$

Theorem 4.3.3. *Let the following conditions hold:*

1. *the principal part of the operator L is independent of t;*
2. *the given functions in (4.3.13) and the domain Ω satisfy the assumptions of Theorem 4.2.2;*
3. *$f(x,0) \neq 0$.*

Then Problem 4.3.3 has at most one solution (u, φ) such that $u \in C^2(\overline{\Omega})$ and $\varphi \in C(\overline{\Omega})$.

The proof of Theorem 4.3.3 is similar to that of Theorem 4.3.2.

A theorem which is analogous to Theorem 4.3.2 holds for the modified Problem 4.3.2 with u_t replaced by $-u_{tt}$ (the elliptic case). The proof of this theorem is essentially similar to that of Theorem 4.3.2.

An inverse problem analogous to Problem 4.3.2 was studied in Bukhgeim (1981) and Klibanov (1981), where data of the form (4.3.2) were specified on the entire boundary of Ω.

3. In this part of the section we assume that the operator L is independent of t and $t \in \mathbb{R}$; furthermore, $|u|, |u_t|, \dots, |u_{x_i x_j}|, \dots, |u_{tt}| < C \exp(kt^2)$ for any fixed x as $|t| \to \infty$, where $C > 0$ and $k > 0$ depend on x and the function u itself.

Problem 4.3.4. Determine the pair of functions (u, φ) from the given relations

$$u_{tt} - Lu = f(x)\varphi(x', t), \tag{4.3.17}$$

$$u(0, 'x, t) = u_{x_1}(0, 'x, t) = 0, \quad x_n \in (-\delta, \delta), \tag{4.3.18}$$

$$u(x', 0, t) = g_1(x', t). \tag{4.3.19}$$

Theorem 4.3.4. *Let the operator L and the solution $u(x,t)$ of the problem (4.3.17) – (4.3.19) satisfy the conditions formulated above, and let $\varphi \in C(\Omega)$, $f(x', 0) \neq 0$.*

Then $g_1 \equiv 0$ implies $(u, \varphi) \equiv 0$ in Ω.

Proof. Set $u^\pm = 2^{-1}(u(x,t) \pm u(x,-t))$, $\varphi^\pm(x',t) = 2^{-1}(\varphi(x,t) \pm \varphi(x', -t))$, $u = u^+ + u^-$, where u^+ is the even component of $u(x,t)$. By the hypothesis of the theorem, if $t > 0$ is sufficiently small, then the following integrals exist:

$$v^+ = \frac{1}{2\sqrt{\pi t}} \int_{-\infty}^{\infty} e^{-\tau^2/4t} u^+(x,\tau)\, d\tau = \frac{1}{\sqrt{\pi t}} \int_0^{\infty} e^{-\tau^2/4t} u^+(x,\tau)\, d\tau,$$

$$\Phi^+ = \frac{1}{\sqrt{\pi t}} \int_0^{\infty} e^{-\tau^2/4t} \varphi^+ \, d\tau. \tag{4.3.20}$$

Hence, since u^+ is even, equalities (4.3.17) – (4.3.19) and the condition $g_1 = 0$ yield

$$v_t^+ - Lv^+ = f(x)\Phi^+, \tag{4.3.21}$$

$$v^+(0, 'x, t) = v_{x_1}^+(0, 'x, t) = 0, \quad x_n \in (-\delta, \delta), \tag{4.3.22}$$

$$v^+(x', 0, t) = 0. \tag{4.3.23}$$

As follows from (4.3.21) – (4.3.23), Theorem 4.3.2 implies $v^+ = \Phi^+ = 0$. Therefore, by the well-known properties of integrals of the form (4.3.20) (see Reznitskaya, 1974), we have $u^+ = \varphi^+ = 0$ in Ω.

Differentiating (4.3.17) and conditions (4.3.18), (4.3.19) with respect to t and putting $u_t^- = z$, we obtain the inverse problem for (z, φ_t^-) analogous to the problem (4.3.21) – (4.3.23), where z and φ_t^- are even functions.

If we set $u^+ = z$, from the above arguments it follows that $z = u_t^- = \varphi_t^- = 0$ in Ω. Since u^- and φ^- are even with respect to t, we have $u^-(x,0) = \varphi^-(x',0) = 0$, which implies that $u^- = \varphi^- = 0$ in Ω. □

4. In the final part of the section, we formulate the theorem on existence "in the small" for Problem 4.3.2.

Theorem 4.3.5. *Let the following conditions hold:*

1. *in some neighbourhood of the origin, the functions a_{ij}, a_l, and f are analytic with respect to all arguments except x_1 and twice continuously differentiable with respect to x_1;*

2. *the principal part of the operator L is independent of x_n;*

3. *$f(x',0,t) \neq 0$.*

Then, provided that $r_0 > 0$ is sufficiently small, Problem 4.3.2 has a unique solution (u, φ) in $\Omega_r = \{(x,t) : 0 < x_1 < r, |'x|^2 + t^2 < r^2\}$ for $0 < r < r_0$ which is analytic with respect to all arguments except x_1 and twice continuously differentiable with respect to x_1.

We briefly outline the proof of Theorem 4.3.5. The original Problem 4.3.2 is reduced to the Cauchy problem for the integro-differential equation (4.3.4) (the problem (4.3.4) – (4.3.5)). In this case, equation (4.3.4) and data (4.3.5) will be inhomogeneous. We may assume that the initial data are zero; otherwise, we can provide this by subtracting an appropriate function. Then the problem (4.3.4) – (4.3.5) is reduced to an equivalent problem for a system of first order integro-differential equations. Finally, we apply the abstract Cauchy – Kovalevskaya Theorem to this equivalent problem (see Nirenberg (1974), Theorem 1.1; it is easy to verify that the said problem satisfies the assumptions of Theorem 1.1 in the scale x_s of vector functions $u(x)$ that are holomorphic and bounded in the domain $D_s = \prod_i \{|x^i| < sR\}$). This leads us to the assertion of Theorem 4.3.5 (see also Ovsyannikov, 1971).

Similar existence theorems hold for other inverse problems discussed above in the second part of Section 4.3.

Remark 4.3.1. The technique of eliminating the unknown coefficient by means of differentiation and introducing an integro-differential equation was first proposed in Beznoshchenko and Prilepko (1977).

Remark 4.3.2. Problems similar to Problem 4.3.1 in the case of small domain Ω and special conditions imposed on the unknown coefficients were studied in Anikonov (1978), Kabanikhin (1978, 1979), Romanov (1984). Uniqueness theorems for a number of inverse problems are presented in Bukhgeim and Klibanov (1981). Some generalizations of these theorems can be found in Isakov (1982) and Khaidarov (1984).

The essential difference between the Problem 4.3.1 and the inverse problems for hyperbolic equations considered in the works mentioned in this remark is that the data in Problem 4.3.1 are given on the plane $x_1 = 0$.

The method presented in Bukhgeim and Klibanov (1981) cannot be applied to Problem 4.3.1, since the Carleman estimates for the Cauchy problem for a hyperbolic equation with data specified on a time like surface in the case of a plane (or a piece of a plane) have not been established so far. The method of proof for Theorem 4.3.1 also differs from the methods presented in Isakov (1982) and Khaidarov (1984).

Problems similar to Problems 4.3.2 and 4.3.3 and analogous problems for elliptic equations were considered "in the small" by Beznoshchenko (1975), Iskenderov (1977), and Yakhno (1982), and "in the large" by Bukhgeim and Klibanov (1981).

The characteristic difference between the formulations of the problems studied in the second and third parts of Section 4.3 and the formulations of the problems in Bukhgeim and Klibanov (1981) is that the data of the direct problem in Section 4.3 are given on a planar piece of the boundary (except for Problem 4.3.1). Moreover, in some inverse problems, the area of this piece can be as small as desired. It is well known that the type of ill-posedness of inverse problems depends on whether the data are given on the whole boundary or on a piece of it. As a rule, they are weakly ill-posed in the case of the whole boundary; otherwise, they are strongly ill-posed.

4.4. INVERSE PROBLEMS WITH CONCENTRATED DATA

We analyze an inverse problem for a hyperbolic equation which is connected with a problem with data containing Dirac's delta function. It is known that the Cauchy problem is equivalent to a certain Goursat problem if the data contain Dirac's delta function (see Romanov, 1984). Depending on the degree of smoothness of the coefficients in the principal part of the equation, the solution to this Goursat problem is nonzero and sufficiently smooth inside the characteristic conoid, and identically equals zero outside the conoid (thus, the solution is discontinuous on the surface of the conoid).

Inverse problems whose data contain Dirac's delta function are associated with problems of integral geometry, in particular with those on ellipsoids (see Romanov, 1972).

The solution to the problem studied in this section vanishes outside the characteristic conoid changing continuously, in contrast to the case of the problem with Dirac's delta functions, where the solution has discontinuity. Such a situation occurs when the delta function in the studied problem is replaced by a smooth function of the "hood" type which is "close" to it. Consider the following equation in $\Omega = \{(x,t) \,:\, x_1 > 0,\, 'x \in \mathbb{R}^{n-1},\, t \in \mathbb{R}\}$:

$$u_{tt} = c^2(x)\Delta_x u + \bar{a}\nabla_{x,t}u + q(x)u, \tag{4.4.1}$$

with the data

$$u(0,'x,t) = u_0, \quad u_{x_1}(0,'x,t) = u_1, \tag{4.4.2}$$

$$u\big|_{t=\tau(x)} = S(x), \tag{4.4.3}$$

where $t = \tau(x)$ is the characteristic conoid of equation (4.4.1) with vertex at the origin (see Romanov, 1984), and $a\nabla_{x,t}u$ is the scalar product of the vector functions $a(x,t)$ and $\nabla_{x,t}u$ in $\mathbb{R}^n$.

Problem 4.4.1. Find a pair of functions (u,q) defined in Ω that satisfies (4.4.1) – (4.4.3), assuming that all the other functions are given.

Let $\mathfrak{M}$ be a set of functions from $C^2(\Omega)$ with the following properties:

1. $|u|$, $|\nabla_{x,t}u|$, $|u_{tt}|$, $|u_{x_ix_j}| \le Ke^{\alpha_0 t}$ as $t \to \infty$ $(i,j = 1,\dots,n)$, the constants K and α_0 depend on $u(x,t)$;

2. $u(x,t) = 0$ for $t < \tau(x) - \delta$, where $\delta > 0$ can be as small as desired;

Set $\mathbb{R}^n_+ = \{x \,:\, x \in \mathbb{R}^n,\, x_1 > 0\}$.

Theorem 4.4.1. *Let the following conditions hold:*

1. $c_{x_1} > 0$ *in* $\overline{\mathbb{R}^n_+}$, *where* $\overline{\mathbb{R}^n_+} = \mathbb{R}^n_+ \cup \{x_1 = 0\}$;

2. $q(x)$ *is known for* $x_1 \in (0,h)$, $h_1 > 0$ *is as small as desired;*

3. $S(x) \ne 0$ *for* $x_1 \le 0$.

Then Problem 4.4.1 has at most one solution $(u, q) \in \mathfrak{M} \times C(\overline{\mathbb{R}}^n_+)$.

Proof. Let $(u^{(1)}, q^{(1)})$ and $(u^{(2)}, q^{(2)})$ be two solutions of Problem 4.4.1 such that $u^{(i)} \in \mathfrak{M}$, $q^{(i)} \in C(\overline{\mathbb{R}}^n_+)$ for $i = 1, 2$. Then (4.4.1) – (4.4.3) imply the following equalities for $u = u^{(2)} - u^{(1)}$ and $\bar{q} = q^{(2)} - q^{(1)}$:

$$u_{tt} = c^2(x)\Delta_x u + a\nabla_{x,t} u + q^{(1)} u + \bar{q} u^{(2)}, \tag{4.4.4}$$

$$u(0, {}'x, t) = u_{x_1}(0, {}'x, t) = 0, \tag{4.4.5}$$

$$u(x, \tau(x)) = 0. \tag{4.4.6}$$

We introduce the function $u = v(x, t)\exp(\lambda_0 t)$, $\lambda_0 > \max\{\alpha_{01}, \alpha_{02}\}$, where α_{01}, α_{02} are such that

$$|u^{(j)}|,\ |\nabla_{x,t} u^{(j)}|,\ |u^{(j)}_{tt}|,\ |u^{(j)}_{x_i x_i}| \le K_j \exp(\alpha_{0j} t),$$

$$j = 1, 2, \quad \alpha_{0j} > 0, \quad i = 1, 2, \dots, n.$$

Using the equality $u_{tt} = (v_{tt} + 2\lambda_0 v_t + \lambda_0^2 v)\exp(\lambda_0 t)$, from (4.4.4) we have

$$v_{tt} = c^2(x)\Delta_x v + a\nabla_x v + (a_{n+1} + 2\lambda_0)v_t + (q^{(1)} + \lambda_0^2)v + \bar{q} v^{(2)}, \tag{4.4.7}$$

and (4.4.5), (4.4.6) imply

$$v(0, {}'x, t) = v_{x_1}(0, {}'x, t) = 0, \tag{4.4.8}$$

$$v(x, \tau(x)) = 0. \tag{4.4.9}$$

Since $q(x)$ in (4.4.1) is known for $0 < x_1 < h$, it follows that $\bar{q} = 0$ for x_1 in the same interval. Therefore, in (4.4.7) we may assume that $v^{(2)} = 0$ for $0 < x_1 < h/2$. Let $\tau(x)$ be a solution to the Cauchy problem for the eikonal equation

$$|\nabla_x \tau| = C^{-2}, \tag{4.4.10}$$

with the data

$$\tau\big|_{x_n=0} = 0, \quad \tau_{x_n}\big|_{x_n=0} = C^{-1}(x', 0). \tag{4.4.11}$$

Introduce the variables

$$x' = y, \quad z = \tau(x), \tag{4.4.12}$$

and the functions

$$w(z,y,t) = v(x,t), \quad \bar{a}(y,z,t) = a(x,t),$$
$$\tilde{q}^{(1)}(z,y) = (\bar{q}^{(1)}(x) + \lambda_0^2), \quad a_{n+1} + 2\lambda_0 = \bar{a}_{n+1}, \tag{4.4.13}$$
$$b(z,y) = c(x), \quad v^{(2)} = w^{(2)}(y,z), \quad \tilde{q}(y,z) = \bar{q}(x).$$

By virtue of the condition $C > 0$, the change of variables (4.4.12) is nondegenerate. It can be verified directly that

$$Lv + \bar{q}v^{(2)} = w_{tt} - w_{zz} + lw + \bar{a}_{n+1}\frac{\partial w}{\partial t} + w^{(2)}\tilde{q} = 0, \tag{4.4.14}$$

where

$$Lv = v_{tt} - c^2\Delta_x v + a\nabla_x v + (a_{n+1} + 2\lambda_0)v_t + (q^{(1)} + \lambda_0^2)v,$$

$$lw = b^2\Delta_y w + \sum_{j=1}^{n-1} p_j \frac{\partial}{\partial y_j}\Big(\frac{\partial w}{\partial z}\Big) + \sum_{j=1}^{n-1}\frac{\partial w}{\partial y_j}\bar{a}_j$$
$$+ \frac{\partial w}{\partial z}\Big(\sum_{j=1}^{n-1}\frac{\partial \tau}{\partial x_j} + b^2\Delta_x\tau + 1\Big) + \tilde{q}^{(1)}w,$$

$$p_j(y,z) = 2b^2\tau_{x_j}, \quad j = 1,2,\dots,n.$$

From (4.4.9) it follows that the derivative of w in the direction of the line $z = t$ is zero because $w(y,z,z) = 0$. Therefore,

$$-w_z = w_t \quad \text{on the plane} \quad t = z. \tag{4.4.15}$$

In order to calculate the expressions

$$\Big(\frac{\partial^2 w}{\partial t^2} - \frac{\partial^2 w}{\partial z^2}\Big)\Big|_{t=z},$$

we represent the wave operator in the form of the product of first order operators:

$$F(y,z,t) \equiv \Big(\frac{\partial^2}{\partial t^2} - \frac{\partial^2}{\partial z^2}\Big)w = \Big(\frac{\partial}{\partial t} + \frac{\partial}{\partial z}\Big)\Big(\frac{\partial}{\partial t} - \frac{\partial}{\partial z}\Big)w.$$

Integrating this equality along the characteristics of the differential operator $\partial/\partial t + \partial/\partial z$ from an arbitrary point (z,t) to the point of intersection with the plane $t = 0$ and using the property $w(y, z, 0) = w_t(y, z, 0) = 0$ ($u_i \in \mathfrak{M}$), we get

$$\Big(\frac{\partial}{\partial t} - \frac{\partial}{\partial z}\Big) w = \int_0^t F(y, \xi - t + z, \xi)\, d\xi.$$

Hence, for $z = t$ we have

$$\int_0^t F(y, \xi, \xi)\, d\xi = \Big(\frac{\partial}{\partial t} - \frac{\partial}{\partial z}\Big)\Big|_{t=z},$$

or

$$\Big(\frac{\partial^2 w}{\partial t^2} - \frac{\partial^2 w}{\partial z^2}\Big)\Big|_{t=z} = \frac{\partial}{\partial t}\Big[\Big(\frac{\partial w}{\partial t} - \frac{\partial w}{\partial z}\Big)\Big|_{t=z}\Big]. \tag{4.4.16}$$

Consequently, by virtue of (4.4.9), (4.4.15), (4.4.16), equality (4.4.14) implies

$$\tilde{q} = [s(y, z)]^{-1}\Big\{-\frac{\partial}{\partial t}\Big[\Big(\frac{\partial w}{\partial t} - \frac{\partial w}{\partial z}\Big)\Big|_{t=z}\Big]$$

$$+ \Big[\sum_{j=1}^{n-1} p_j \frac{\partial^2 w}{\partial y_j \partial t} + \frac{\partial w}{\partial t} A\Big]\Big|_{t=z}\Big\} \exp(-\lambda_0 z), \tag{4.4.17}$$

where

$$A = \sum_{j=1}^{n} \frac{\partial \tau}{\partial x_j} + b^2 \Delta \tau - \bar{a}_{n+1} + 1.$$

Substitute the expression (4.4.17) for $\tilde{q}$ in (4.4.14). As a result, we obtain the Cauchy problem for the equation

$$w_{tt} - w_{zz} + lw + w^{(2)}[s(y, z)]^{-1} \exp(-\lambda_0 z)\Big\{-\frac{\partial}{\partial t}\Big[\Big(\frac{\partial w}{\partial t} - \frac{\partial w}{\partial z}\Big)\Big|_{t=z}\Big]$$

$$+ \Big[\sum_{j=1}^{n-1} p_j \frac{\partial^2 w}{\partial y_j \partial t} + A\frac{\partial w}{\partial t}\Big]\Big|_{t=z}\Big\} = 0 \tag{4.4.18}$$

with the data

$$w(0, y_2, \dots, y_{n-1}, z) = w_{y_1}(0, y_2, \dots, y_{n-1}, z) = 0. \tag{4.4.19}$$

Condition (4.4.19) follows from (4.4.8). Applying the Fourier transform with respect to t to the problem (4.4.18) – (4.4.19) and using the formulas

$$\frac{\partial}{\partial t}\Big[\Big(\frac{\partial w}{\partial t} - \frac{\partial w}{\partial z}\Big)\Big|_{t=z}\Big] = -\int_{\mathbb{R}} e^{i\xi z}\Big(i\xi\frac{\partial}{\partial z} + \xi^2\Big)\hat{w}(y, z, \xi)\, d\xi,$$

$$\Big[\sum_{j=1}^{n-1} p_j \frac{\partial^2 w}{\partial y_j \partial t} + A\frac{\partial w}{\partial t}\Big]\Big|_{t=z} = \sum_{j=1}^{n-1} \int_{\mathbb{R}} \Big(p_j \frac{\partial \hat{w}}{\partial y_j} + A\hat{w}\Big)(\mathrm{i}\xi)\mathrm{e}^{\mathrm{i}\xi z}\,\mathrm{d}\xi,$$

we obtain

$$-\xi^2\hat{w} - \hat{w}_{zz} + l\hat{w} + \widehat{\bar{a}_{n+1}w_t}$$
$$+ \hat{w}^{(2)}\exp(\lambda_0 z)[s(y,z)]^{-1} \int_{\mathbb{R}} \mathrm{e}^{\mathrm{i}\xi z}\Big[\mathrm{i}\xi\frac{\partial \hat{w}}{\partial z} + \xi^2\hat{w} + (\mathrm{i}\xi)A\hat{w}$$
$$+ \sum_{j=1}^{n-1} p_j \frac{\partial \hat{w}(y,z,\xi)}{\partial y_j}(\mathrm{i}\xi)\Big]\,\mathrm{d}\xi = 0, \tag{4.4.20}$$

$$\hat{w}(0,y_2,\dots,y_{n-1},z,\xi) = \hat{w}_{y_1}(0,y_2,\dots,y_{n-1},z,\xi) = 0, \tag{4.4.21}$$

where $\hat{w}$ and $\hat{w}^{(2)}$ are the Fourier transforms of w and $w^{(2)}$, respectively.

We now return to the variables x and the functions v, q, c, a in the problem (4.4.20), (4.4.21). It is possible because the change of variables (4.4.12), (4.4.13) is reversible. As a result, using the property $\partial x_i/\partial z = 0$ $(i = 1,2,\dots,n-1)$, we have

$$-\xi^2\hat{v} - c^2(x)\Delta_x\hat{v} + a\,\nabla_x\hat{v} + \widehat{(a_{n+1} + 2\lambda_0)v_t}$$
$$+ (q^{(1)} + \lambda_0^2)\hat{v} + \hat{v}^{(2)}\exp(\lambda_0\tau(x))[s(x',\tau(x)]^{-1} \times$$
$$\times \int_{\mathbb{R}} \exp(\mathrm{i}\xi\tau(x))\Big[\mathrm{i}\xi\frac{\partial \hat{v}}{\partial z} + \xi^2 v$$
$$+ \sum_{j=1}^{n-1} c^2(x)\tau_{x_j}\frac{\partial \hat{v}}{\partial x_j}(\mathrm{i}\xi) + (\mathrm{i}\xi)A'\hat{v}\Big]\,\mathrm{d}\xi = 0, \tag{4.4.22}$$

where

$$A' = \sum_{j=1}^{n-1} \frac{\partial \tau}{\partial x_j} + c^2\Delta\tau - a_{n+1} + 1,$$

and $\hat{v}$ is the Fourier transform of v with respect to t. Furthermore, from (4.4.21) it follows that

$$\hat{v}(0,{}'x,t) = \hat{v}_{x_1}(0,{}'x,t) = 0. \tag{4.4.23}$$

Thus, we arrive at the problem (4.4.22) – (4.4.23), which is similar to the problem (4.2.4) – (4.2.5) considered in Section 4.2 in the case where $m = 1$,

$a_j = b_j = 0$, $j = 1, 2, \dots, n$. The only difference is that equation (4.4.22) contains the terms of the new type:

$$\int_{\mathbb{R}} \exp(\mathrm{i}\xi\tau(x))(\mathrm{i}\xi)\frac{\partial \hat{v}}{\partial z}\,\mathrm{d}\xi, \qquad \int_{\mathbb{R}} \frac{\partial \hat{v}(x,\xi)}{\partial x_j}\exp(\mathrm{i}\xi\tau(x))(\mathrm{i}\xi)\,\mathrm{d}\xi \tag{4.4.24}$$

with known multipliers.

It is clear from the proof of the uniqueness of solution to the problem (4.2.4) – (4.2.5) that a similar result holds in the case where equation (4.2.4) contains the terms of the form (4.4.24) (see estimate (4.2.21)). Therefore, from (4.4.22) – (4.4.23) it follows that $\hat{v} = 0$, which implies $\bar{q} = 0$ because of (4.4.7) and the relation $v^{(2)}(x, \tau(x)) = s(x) \neq 0$. This completes the proof of the theorem. □

Appendix A.

Here we present the most important kinetic equations.

Kinetic equations characterize continuous motion of a substance and are fundamental equations in science.

They are widely used for qualitative and quantitative description of physical, chemical, biological, and other kinds of processes on a microscopic scale. Kinetic equations are often referred to as master equations since they play an important role in the theory of substance motion under the action of forces, in particular, irreversible processes. Kinetic equations are mostly integro-differential equations that are often nonlinear with respect to one or several distribution functions $F(x, p, t)$ characterizing the dynamic states of a substance depending on the space coordinates x, momentum p, and time t. Quantum terms appear in kinetic equations if strong interactions are present. The examples of kinetic equations are the Liouville equation, the Boltzmann equation, the Vlasov equation, the chain of Bogolyubov equations, the Fokker-Planck equation, various equations of radiation transfer, quantum equations for the density matrix and Wigner functions.

Information on kinetic equations and their applications can be found in Case and Zweifel (1967), Klimantovich (1982), Liboff (1979), Polyachenko and Fridman (1976), Silin (1971), and Cercignani (1975). See also the bibliography therein. We now give examples of kinetic equations. In the sequel, the distribution function $F(x, p, t)$ is assumed to be sufficiently differentiable and decreasing rapidly enough with respect to p; $x \in D$, $p \in \mathbb{R}^n$, $t \in \mathbb{R}^1$, D is a domain with smooth boundary ∂D. The product $F(x, p, t)\Delta x \Delta p$ represents the number of particles in the unit element of phase volume at the moment t.

1. The Liouville equation can be represented in the form

$$\frac{\partial F}{\partial t} + \{F, H\} = 0, \tag{1}$$

where $F(x, p, t)$ is the distribution function for a certain kind of particles, $x \in D$, $p \in \mathbb{R}^n$, $t \in \mathbb{R}^1$, $H = H(x, p, t)$ is the Hamiltonian function,

$$\{F, H\} = \sum_{j=1}^{n} \Big(\frac{\partial F}{\partial x_j} \frac{\partial H}{\partial p_j} - \frac{\partial F}{\partial p_j} \frac{\partial H}{\partial x_j} \Big)$$

is the Poisson bracket of F and H.

The Liouville equation characterizes the continuity of the motion of a substance with phase volume conservation. The Hamiltonian function in (1) often has the form

$$H = \frac{1}{2}|p|^2 - \Phi(x, t).$$

In this case, the Liouville equation can be written as

$$\frac{\partial F}{\partial t} + \sum_{j=1}^{n} \Big(\frac{\partial F}{\partial x_j} p_j + \frac{\partial F}{\partial p_j} \frac{\partial \Phi}{\partial x_j} \Big) = 0.$$

2. The Boltzmann equation has the form

$$\frac{\partial F}{\partial t} + \sum_{j=1}^{n} \frac{\partial F}{\partial x_j} p_j = stF.$$

Here stF is the so-called collision integral, which characterizes the variation of the distribution function $F(x, p, t)$ with pairwise collisions taken into account. In the case of the Boltzmann equation, stF is represented in the form

$$stF = \int_{\mathbb{R}^n} \int_{\mathbb{R}^n} \int_{\mathbb{R}^n} w(p, p_1; p', p_1')[F(x, p, t)F(x, p_1, t) - F(x, p', t)F(x, p_1', t)]\, dp_1\, dp'\, dp_1',$$

where $w(p, p_1; p', p_1')$ is a function characterizing the probability of the collision of two particles with momenta p and p_1 with their transition to the state p', p_1', such that

$$w(p, p_1; p', p_1') = w(p', p_1'; p, p_1).$$

This equality describes the detailed equilibrium: the collision process is balanced by the reverse process.

The equation of the Boltzmann type is widely used for the qualitative and quantitative description of processes in gas dynamics.

3. Quantum kinetic equations are integro-differential equations of the form

$$\frac{\partial F}{\partial t} + \{F, H\} + \sigma F$$
$$= \frac{1}{(2\pi)^n h} \int_{\mathbb{R}^{2n}} [\Phi(x - \frac{1}{2}hy, t) - \Phi(x + \frac{1}{2}hy, t)] \mathrm{e}^{-\mathrm{i}y(p-p')} F(x, p', t) \, \mathrm{d}p' \, \mathrm{d}y$$
$$+ stF + q(x, p, t). \tag{2}$$

Here $F(x, p, t)$ is the quantum distribution function, H is the Hamiltonian function, Φ is the mean potential, $\sigma(x, p, t)$ is the absorption, h is Planck's constant, stF is the collision integral, $q(x, p, t)$ is the function describing the sources. The existence of the integral

$$\int_{\mathbb{R}^n} [\Phi] \exp[-\mathrm{i}y(p - p')] \, \mathrm{d}p' \, \mathrm{d}y$$

implies the quantum character of the equation. If $\sigma = 0$, $stF = 0$, $q = 0$, and $H = |p|^2/2 + \Phi(x, t)$, then the quantum distribution function is related to the wave function $\psi(x, t)$ by the equality

$$F(x, p, t) = \frac{1}{(2\pi)^n} \int_{\mathbb{R}^n} \psi^* \Big(x - \frac{1}{2}hy, t\Big) \mathrm{e}^{\mathrm{i}yp} \, \mathrm{d}y.$$

The wave function $\psi(x, t)$ is the solution of the Schrödinger equation:

$$\mathrm{i}\frac{\partial \psi}{\partial t} = -\frac{1}{2h^2}\Delta\psi + \Phi\psi.$$

By expanding in powers of h, the quantum kinetic equation can be formally represented in the form of an equation of infinite order for p:

$$\frac{\partial F}{\partial t} + \{F, H\} + \sigma F = \sum C_\alpha h^{2\alpha} D_x^{2\alpha+1} \Phi D_p^{2\alpha+1} F + stF + q,$$

where C_α are constants, D_x and D_p are the differential operators with respect to x and p, respectively.

The foregoing equation is used to obtain various classical approximations, such as the following (only the terms with $\alpha = 0$ are taken into account):

$$\frac{\partial F}{\partial t} + \{F, H\} + \sigma F = \sum \frac{\partial \Phi}{\partial x_j} \partial F \partial p_j + stF + q.$$

If $H = |p|^2/2$, $\partial\sigma/\partial p_j = 0$, and $stF = 0$, then the procedure of Fourier transform of equation (2) with respect to p leads to the following second order equation for the function $w(x, y, t) = \dfrac{1}{(2\pi)^n} \displaystyle\int_{\mathbb{R}^n} F e^{-\mathrm{i}yp}\, \mathrm{d}p$:

$$\frac{\partial w}{\partial t} - \mathrm{i} \sum_{i=1}^{n} \frac{\partial^2 w}{\partial x_i^2} + \sigma w$$
$$= \Big[\Phi\Big(x - \frac{1}{2}hy, t\Big) - \Phi\Big(x + \frac{1}{2}hy, t\Big)\Big] w + \hat{q}.$$

4. Linear equations of transport theory are called balance equations:

$$\frac{\partial F}{\partial t} + \sum_{i=1}^{n} v_i \frac{\partial F}{\partial x_i} + |\vec{v}|\sigma(x, v, t) F$$
$$= \int_{\mathbb{R}^n} w(v', v, x, t) F(x, v', t)\, \mathrm{d}v' + q(x, v, t). \tag{3}$$

Here $F \Delta x \Delta v$ represents the number of particles, for example, neutrons, in the unit volume element $\Delta x \Delta v$ at the moment t; w is the scattering indicatrix, $q(x, v, t)$ is the function characterizing the sources, $1/\sigma$ is the mean free path length of the particles, $|\vec{v}|$ is the absolute value of the vector $\vec{v} = (v_1, v_2, \dots, v_n)$.

Linear equations of the form (3) are derived from nonlinear equations of the Boltzmann type. The absolute value of the velocity $\vec{v}$ in (3) is often assumed to be constant. In this case, equation (3) is written in the following form after the appropriate modifications:

$$\frac{\partial F}{\partial t} + |\vec{v}|\, \Omega \operatorname{grad}_x F + |\vec{v}| \sigma F$$
$$= \int_{\omega_n} w(x, \Omega, \Omega', t) F\, \mathrm{d}\Omega' + q(x, \Omega', t), \tag{4}$$

where $|\Omega'| = 1$, $|\Omega| = 1$, and ω_n is the unit sphere.

Equation (4) is called the nonstationary linear equation of transport theory, in particular, neutron transport theory (see Case and Zweifel, 1967).

5. The Vlasov system. This system of kinetic equations describes the interaction of charged particles in the electromagnetic field. It has the following form:

$$\frac{\partial F_e}{\partial t} + \vec{v}\frac{\partial F_e}{\partial x} - e\Big(\vec{E} + \frac{1}{c}[\vec{v}\vec{B}]\Big)\frac{\partial F_e}{\partial p} = 0$$

$$\frac{\partial F_i}{\partial t} + \vec{v}\frac{\partial F_i}{\partial x} + ze\Big(\vec{E} + \frac{1}{c}[\vec{v}\vec{B}]\Big)\frac{\partial F_e}{\partial p} = 0$$

$$\operatorname{rot}\vec{E} = -\frac{1}{c}\frac{\partial \vec{B}}{\partial t}, \quad \operatorname{rot}\vec{B} = \frac{1}{c}\frac{\partial \vec{E}}{\partial t} + 4\pi c\vec{J}. \tag{5}$$

$$\operatorname{div}\vec{B} = 0, \quad \operatorname{div}\vec{E} = 4\pi\rho,$$

$$\rho = e\int_{\mathbb{R}^n}(zF_i - F_e)\,\mathrm{d}\vec{p}, \quad \gamma = e\int_{\mathbb{R}^n}(zF_i - F_e)\vec{v}\,\mathrm{d}\vec{p}.$$

Here F_e and F_i are the distribution functions for electrons and ions, respectively; ρ and γ are the charge and current densities, respectively; $\vec{E}$ and $\vec{B}$ are the electric and magnetic field strengths, e is the electron charge and z is the ion charge.

The system (5) is a nonlinear system of integro-differential equations for F_e, F_i, $\vec{E}$, H.

6. The Fokker-Planck equations are usually used for describing slow processes of transport and diffusion of a small admixture of a heavy gas in a light gas. These are linear integro-differential equations. The general form of such equations is

$$\frac{\partial F}{\partial t} + \{F, H\} = \int_{\mathbb{R}^n}[w(p+q,q)F(p,p+q,t) - w(p,q)F(p,q,t)]\,\mathrm{d}q. \tag{6}$$

Here $w(p,q)$ is the probability of a heavy particle to change its momentum. Using the Taylor series expansion of the integrand in (6), one can obtain the

Fokker-Planck kinetic second order differential equations:

$$\frac{\partial F}{\partial t} + \{F, H\} = \sum_{j=1}^{n} \frac{\partial}{\partial p_j} A_j F + \sum_{j,k=1}^{n} \frac{\partial}{\partial p_j} \frac{\partial}{\partial p_k} (B_{kj} F) = 0,$$

$$A_j = \int_{\mathbb{R}^n} q_j w \, \mathrm{d}q, \quad B_{kj} = \int_{\mathbb{R}^n} q_j q_k w \, \mathrm{d}q,$$

7. The chain of Bogolyubov equations. This is a system of equations represented by a chain of kinetic equations for k-partial distribution functions,

$$F_k(x^1, x^2, \dots, x^k, p^1, \dots, p^k, t) = v^k \int F_N \, \mathrm{d}x^{k+1} \dots \mathrm{d}x^N \, \mathrm{d}p^{k+1} \dots \mathrm{d}p^N,$$

where $k = 1, 2, \dots, N$ and F_N is the solution of the Liouville equation

$$\frac{\partial F_N}{\partial t} + \{F_N, H_N\} = 0,$$

v is the volume of the phase-space domain under study.

The chain of Bogolyubov equations is written in the following form:

$$\frac{\partial F_k}{\partial t} + \{F_k, H_k\} = \int_{\mathbb{R}^{2n}} \{\Phi_k, F_{k+1}\} \, \mathrm{d}x^{k+1} \, \mathrm{d}p^{k+1},$$
$$k = 1, 2, \dots, N$$

Here H_k is the Hamiltonian function and Φ_k is the potential.

8. Kinetic equations in problems of integral geometry.

Let $\lambda(x,t)$ be a finite continuously differentiable function of $x \in \mathbb{R}^n$ and $t \in \mathbb{R}$. By $\Gamma(x,p,t)$ we denote the ray $(x + p(\tau - t), \tau)$, $t \le \tau < \infty$, with vertex at the point (x,t) and direction p.

We set

$$F(x,p,t) = \int_t^{\infty} \lambda(x + p(\tau - t), \tau) \, \mathrm{d}\tau.$$

Obviously, $F(x,p,t)$ is the integral of λ along the ray $\Gamma(x,p,t)$. From the definition of $F(x,p,t)$ it is clear that it satisfies the kinetic equation

$$\frac{\partial F}{\partial t} + \sum_{j=1}^{n} \frac{\partial F}{\partial x_j} p_j + \lambda = 0.$$

Note that λ can depend also on the direction p. Replace the rays $\Gamma(x,p,t)$ by the curves satisfying, for example, the Hamiltonian system with $H(x,p)$. Then, for the integral $F(x,p,t)$ over such curves with vertex at (x,t) and direction p, we have the Liouville equation with right-hand side λ:

$$\frac{\partial F}{\partial t} + \{F, H\} = \lambda.$$

In the case of natural parametric equations of lines and curves (with τ being the arc length), the corresponding kinetic equations for the integrals of λ have the "stationary" form

$$\sum_{j=1}^{n} \frac{\partial F}{\partial x_j} \Omega_j = \lambda, \quad |\vec{\Omega}| = 1.$$

In the more general case of arbitrary curves, we have

$$\sum_{j=1}^{n} \Big(\frac{\partial F}{\partial x_j} \Omega_j + \frac{\partial F}{\partial \varphi_j} \Big) K_j(x, \vec{\Omega}) = \lambda.$$

In particular, if $n = 2$, $x_1 = x$, $x_2 = y$, then

$$\frac{\partial F}{\partial x} \cos\varphi + \frac{\partial F}{\partial y} \sin\varphi + K(x,y,\varphi) \frac{\partial F}{\partial \varphi} = \lambda$$

for $F(x,y,\varphi) = \int_{\Gamma(x,y,\varphi)} \lambda \, \mathrm{d}s$, where $\Gamma(x,y,\varphi)$ is the part of the curve with initial point (x,y), the tangent angle and curvature at (x,y) being φ and $K(x,y,\varphi)$, respectively.

If the curves Γ are geodesics of the metric $\mathrm{d}s^2 = a(\mathrm{d}x^2 + \mathrm{d}y^2)$, then

$$K = -\Big(\frac{a_x}{a} \sin\varphi - \frac{a_y}{a} \cos\varphi \Big).$$

For $a = \lambda$ we have the equation of inverse kinematic problem of seismology

$$\frac{\partial F}{\partial x} \cos\varphi + \frac{\partial F}{\partial y} \sin\varphi - \Big(\frac{\lambda_x}{\lambda} \sin\varphi - \frac{\lambda_y}{\lambda} \cos\varphi \Big) \frac{\partial F}{\partial \varphi} = \lambda.$$

In the problems of integral geometry, F and λ are the unknown functions to be determined. The data are represented by the trace of F on a subset of the space of variables (x,p,t) with dimension less that $2n+1$.

Bibliography

Alexeev, B. V. (1982). *Mathematical Kinetics of Reacting Gases.* Nauka, Moscow (in Russian).

Amirov, A. Kh. (1977). A problem of integral geometry and a mixed problem. In: *Well-Posedness of Problems of Mathematical Physics*, 3–9, Novosibirsk (in Russian).

Amirov, A. Kh. (1983). A class of multidimensional inverse problems. *Dokl. Akad. Nauk SSSR*, **272**, No. 3, 265–267 (in Russian).

Amirov, A. Kh. (1984a). An inverse problem for the kinetic equation. In: *Ill-Posed Problems of Mathematical Physics and Analysis*, 4–8, Nauka, Novosibirsk (in Russian).

Amirov, A. Kh. (1984b). A Class of Inverse Problems for Kinetic Equations. Preprint No. 493, Computing Center, Siberian Branch of the USSR Acad. Sci. (in Russian).

Amirov, A. Kh. (1985). A class of inverse problems for kinetic equations. *Dokl. Akad. Nauk SSSR*, **283**, No. 4, 780–783 (in Russian).

Amirov, A. Kh. (1986a). Existence and uniqueness theorems for the solution of an inverse problem for the transport equation. *Sibirsk. Mat. Zh.*, **27**, No. 6, 3–20 (in Russian).

Amirov, A. Kh. (1986b). Solvability of inverse problems for a second-order equation. *Funktsional. Anal. i Prilozhen.*, **20**, No. 3, 80–81 (in Russian).

Amirov, A. Kh. (1986c). The question of the solvability of inverse problems. *Dokl. Akad. Nauk SSSR*, **290**, No. 2, 268–270 (in Russian).

Amirov, A. Kh. (1986d). *The Existence and Uniqueness of a Solution to an Inverse Problem for the Kinetic Equation.* Preprint No. 692, Computing Center, Siberian Branch of the USSR Acad. Sci. (in Russian).

Amirov, A. Kh. (1986e). On the solvability of inverse problems for second order equations. In: *Questions of Well-Posedness and Methods for Studying Inverse Problems*, 3–19, Novosibirsk (in Russian).

Amirov, A. Kh. (1987a). A class of inverse problems for a special kinetic equation. *Dokl. Akad. Nauk SSSR*, **295**, No. 2, 265–267 (in Russian).

Amirov, A. Kh. (1987b). Uniqueness of the solution of an inverse problem for the kinetic equation. *Sibirsk. Mat. Zh.* **28**, No. 5, 3–5 (in Russian).

Amirov, A. Kh. (1987c). On a problem of the solvability of inverse problems. *Sibirsk. Mat. Zh.* **28**, No. 6, 3–11 (in Russian).

Amirov, A. Kh. (1987d). A class of inverse problems for a special kinetic equation. In: *Methods for Studying Inverse and Ill-Posed Problems*, 4–21, Novosibirsk (in Russian).

Amirov, A. Kh. and Pashaev, R. T. (1992). An inverse problem for the wave equation, *Soviet Math. Dokl.*, **44**, No. 1, 121–122.

Anikonov, D. S. (1984a). Multidimensional inverse problems for the transport equation. *Differents. Uravn.* **20**, No. 5, 817–824 (in Russian).

Anikonov, D. S. (1984b). Uniqueness of the simultaneous determination of two coefficients of the transport equation. *Dokl. Akad. Nauk SSSR*, **277**, No. 4, 777–779 (in Russian).

Anikonov, D. S. (1985). Solution of the tomographic problem for a source of special type. In: *Linear and Nonlinear Problems of Computational Tomography*, 3–10, Novosibirsk (in Russian).

Anikonov, Yu. E. (1976). The solvability of a certain problem of integral geometry. *Mat. Sb.*, **101(143)**, No. 2, 271–279 (in Russian).

Anikonov, Yu. E. (1978). *Some Methods of Studying Multidimensional Inverse Problems for Differential Equations.* Nauka, Novosibirsk (in Russian).

Anikonov, Yu. E. (1987). New mathematical models and inverse problems. In: *Mathematical Modelling in Geophysics*, 3–42, Nauka, Novosibirsk (in Russian).

Anikonov, Yu. E. (1997). *Multidimensional Inverse and Ill-Posed Problems for Differential Equations.* VSP, Utrecht.

Anikonov, Yu. E. and Amirov, A. Kh. (1983). A uniqueness theorem for the solution of an inverse problem for the kinetic equation. *Dokl. Akad. Nauk SSSR*, **272**, No. 6, 1292–1293 (in Russian).

Anikonov, Yu. E. and Bondarenko, A. N. (1982). An inverse problem for the Vlasov equation. *Dokl. Akad. Nauk SSSR*, **265**, No. 5, 1037–1039 (in Russian).

Anikonov, Yu. E. and Bondarenko, A. N. (1984). Multidimensional inverse problems for kinetic equations. *Dokl. Akad. Nauk SSSR*, **277**, No. 4, 779–781 (in Russian).

Belishev, M. I. (1988). An approach to multidimensional inverse problems for the wave equation. *Soviet Math. Dokl.*, **36**, No. 3, 481–484.

Bernshtein, I. N. and Gerver, M. L. (1978). A problem of integral geometry for a family of geodesics and the inverse kinematic problem of seismology. *Dokl. Akad. Nauk SSSR*, **243**, No. 2, 302–305 (in Russian).

Bernshtein, I. N. and Gerver, M. L. (1980). Conditions for distinguishability of metrics on the basis of time-distance plots. In: *Methods and Algorithms of Seismology Data Interpretation. Computational Seismology*, **13**, 50–73, Nauka, Moscow (in Russian).

Beznoshchenko, N. Ya. (1975). Determining the coefficients of lower-order terms in a parabolic equation. *Sibirsk. Mat. Zh.*, **16**, No. 3, 473–482 (in Russian).

Beznoshchenko, N. Ya. and Prilepko, A. I. (1977). Inverse problems for parabolic equations. In: *Problems of Mathematical Physics and Computational Mathematics*, 51–63, Nauka, Moscow (in Russian).

Blagoveshchenskii, A. S. (1990). On the reconstruction of functions from the integrals along linear manifolds. *Math. Notes*, **39**, 841–849.

Bukhgeim, A. L. (1972a). A certain class of Volterra operator equations of the first kind. *Funktsional. Analiz i Prilozhen.*, **6**, No. 1, 1–9 (in Russian).

Bukhgeim, A. L. (1972b). Certain problems of integral geometry. *Sibirsk. Mat. Zh.*, **13**, No. 1, 34–42 (in Russian).

Bukhgeim, A. L. (1981). Carleman estimates for Volterra operators and uniqueness in inverse problems. In: *Nonclassical Problems of Mathematical Physics*, 56–64, Novosibirsk (in Russian).

Bukhgeim, A. L. (1983). An algorithm of solving the inverse kinematic problem of seismology. In: *Computational Methods in Seismic Research*, 152–155, Nauka, Novosibirsk (in Russian).

Bukhgeim, A. L. and Klibanov, M. V. (1981). Uniqueness in the large of a class of multidimensional inverse problems. *Dokl. Akad. Nauk SSSR*, **260**, No. 2, 269–272 (in Russian).

Carleman, T. (1939). Sur un problème d'unicité pur les systèmes d'équations aux dérivées à deux variables indépendantes. *Ark. Mat., Astr. Fys.*, **26 B**, No. 17, 1–9.

Case, K. and Zweifel, P. (1967). *Linear Transport Theory.* Addison-Wesley Publishing Co., Reading.

Cercignani, C. (1975). *Theory and Application of the Boltzmann Equation.* Scottish Academic Press, Edinburgh.

Courant, R. (1964). *Partial Differential Equations.* Mir, Moscow (in Russian).

Dubrovin, B. A., Novikov, S. P., and Fomenko, A. T. (1979). *Modern Geometry.* Nauka, Moscow (in Russian).

Dunford, N. and Schwartz, J. T. (1958). *Linear Operators. Vol. II.* John Wiley & Sons, New York.

Gel'fand, I. M. and Graev, M. I. (1959). Geometry of homogeneous spaces, representations of groups in homogeneous spaces and related questions of integral geometry. In: *Trudy Moskov. Mat. Obsh.*, **8**, 321–390 (in Russian).

Gel'fand, I. M., Graev, M. I., and Vilenkin, N. Ya. (1962). Integral geometry and the related questions of the theory of representations. *Ser.: Generalized Functions*, **5**, Fizmatgiz, Moscow (in Russian).

Gerver, M. L., and Markushevich, V. M. (1967). Studying the propagation of a seismic wave on the basis of the time-distance plot. In: *Methods and Programs for Seismic Observations Analysis*, **3**, 3–51, Nauka, Moscow (in Russian).

Golubyatnikov, V. P. (1995). On uniqueness of reconstruction of the form of convex and visible bodies from their projections. *Amer. Math. Soc. Transl.*, Ser. 2, , No. 163, 35–45.

Gromoll, D., Klingenberg, W., and Meyer, W. (1975). *Riemannsche Geometrie im Grossen.* Lecture Notes in Math., **55**, Springer-Verlag, Berlin.

Hamaker, C., Smith, K., Solmon D., and Wagner S. (1980). The divergent beam X-ray transform. *Rocky Mountain J. Math.*, **10**, No. 1, 253–283.

Helgason, S. (1980). *The Radon Transform.* Birkhauser, Boston.

Herman, G. (1983). *Image Reconstruction from Projections. Fundamentals of Reconstructive Tomography.* Mir, Moscow (in Russian).

Introduction to Modern Tomography (1983). (Ed. by K. S. Ternovii and M. V. Sin'kov) Naukova Dumka, Kiev (in Russian).

Iordanskii, S. V. (1961). The Cauchy problem for the kinetic equation of plasma. *Trudy Mat. Inst. Steklov*, **60**, 181–194 (in Russian).

Isakov, V. M. (1982). A class of inverse problems for parabolic equations. *Dokl. Akad. Nauk SSSR*, **263**, No. 6, 1296–1299 (in Russian).

Iskenderov, A. D. (1976). Some inverse problems of determining filtration, thermal, and physical characteristics. In: *Nonclassical Methods in Geophysics*, 54–63, Novosibirsk (in Russian).

Iskenderov, A. D. (1977). Some inverse problems of determining the right-hand sides of differential equations. *Izv. Akad. Nauk Azerbaidzhan. SSR, Ser. Fiz.-Tekhn. Mat. Nauk*, , No. 2, 58–63 (in Russian).

Ivanov, V. K., Vasin, V. V., and Tanana, V. P. (1978). *Theory of Linear Ill-Posed Problems and its Applications.* Nauka, Moscow (in Russian).

John, F. (1935). Abhändigkeiten zwischen den Fläschenintegralen einer stetigen Funktion. *Math. Ann.*, , No. 111, 541–559.

John, F. (1943). Bestimmung einer Funktion aus ihren Integralen über gewisse Mannigfaltigkeiten. *Math. Ann.*, , No. 109, 488–520.

John, F. (1955). *Plane Waves and Spherical Means.* Interscience Publishers, New York.

Kabanikhin, S. I. (1978). *On the Uniqueness of Determination of the Coefficients of a Hyperbolic Equation.* Preprint No. 91, Computing Center, Siberian Branch of the USSR Acad. Sci. (in Russian).

Kabanikhin, S. I. (1979). Application of energy inequalities to an inverse problem for a hyperbolic equation. *Differents. Uravn.*, **15**, No. 1, 61–67.

Kabanikhin, S. I. and Boboev, K. (1981). The inverse problem for the P_n-approximation of the kinetic transport equation. In: *The Study of Well-Posedness of Inverse Problems and Certain Operator Equations*, 44–45, Novosibirsk (in Russian).

Khaidarov, A. (1984). Carleman estimates and inverse problems for hyperbolic equations. *Dokl. Akad. Nauk SSSR*, **279**, No. 4, 817–820 (in Russian).

Kireitov, V. R. (1975). The problem of determining an optical surface from its representations. *Funktsional. Anal. i Prilozhen.*, **10**, No. 3, 45–54 (in Russian).

Kireitov, V. R. (1983). *Inverse Problems of Photometry.* Computing Center of the USSR Acad. Sci., Novosibirsk (in Russian).

Klibanov, M. V. (1976a). A certain problem of integral geometry, and the inverse problem for a parabolic equation. *Sibirsk. Mat. Zh.*, **17**, No. 1, 75–84 (in Russian).

Klibanov, M. V. (1976b). The inverse problem for a parabolic equation, and a certain problem of integral geometry. *Sibirsk. Mat. Zh.*, **17**, No. 3, 564–569 (in Russian).

Klibanov, M. V. (1981). Uniqueness of solutions "in the large" of certain multidimensional inverse problems. In: *Nonclassical Problems of Mathematical Physics*, 101–114, Computing Centre, Siberian Branch of the USSR Acad. Sci., Novosibirsk (in Russian).

Klimantovich, Yu. L. (1982). *Static Physics.* Nauka, Moscow (in Russian).

Krein, S. G. (1967). *Linear Differential Equations in a Banach Space.* Nauka, Moscow (in Russian).

Krein, S. G. and Prozorovskaya, O. I. (1960). Analytic semi-groups and incorrect problems for evolutionary equations. *Soviet Math. Dokl.*, **1**, 841–844.

Kostelyanets, P. O. and Reshetnyak, Yu. G. (1954). Defining an absolutely additive function by specifying its values on half-spaces. *Uspekhi Mat. Nauk*, **9**, No. 3, 135–140 (in Russian).

Lavrent'ev, M. M. (1967). *Some Improperly Posed Problems of Mathematical Physics.* Springer-Verlag, New York.

Lavrent'ev, M. M. and Amirov, A. Kh. (1976). A study of the conditional correctness of the Cauchy problem for a certain equation of evolutionary type. *Dokl. Akad. Nauk SSSR*, **227**, No. 5, 1053–1055 (in Russian).

Lavrent'ev, M. M. and Amirov, A. Kh. (1977a). A certain conditionally well-posed problem. *Dokl. Akad. Nauk SSSR*, **232**, No. 4, 773–776 (in Russian).

Lavrent'ev, M. M. and Amirov, A. Kh. (1977b). Conditional correctness of the Cauchy problem. *Fuktsional. Analiz i Prilozhen.*, **11**, No. 2, 80–82 (in Russian).

Lavrent'ev, M. M. and Amirov, A. Kh. (1977c). A study of conditions for the correctness of the Cauchy problem for a certain equation. *Sibirsk. Mat. Zh.*, **18**, No. 5, 1065–1072 (in Russian).

Lavrent'ev, M. M. and Anikonov, Yu. E. (1967). A certain class of problems of integral geometry. *Dokl. Akad. Nauk SSSR*, **176**, No. 5, 1240–1241 (in Russian).

Lavrent'ev, M. M. and Bukhgeim, A. L. (1973). A certain class of problems of integral geometry. *Dokl. Akad. Nauk SSSR*, **211**, No. 1, 38–39 (in Russian).

Lavrent'ev, M. M., Derevtsov, E. Yu., and Sharafutdinov, V. A. (1981). Determination of an optical body that is in a homogeneous medium from its projections. *Dokl. Akad. Nauk SSSR*, **260**, No. 4, 799–803 (in Russian).

Lavrent'ev, M. M. and Romanov, V. G. (1966). Three linearized inverse problems for hyperbolic equations. *Dokl. Akad. Nauk SSSR*, **171**, No. 6, 1279–1281 (in Russian).

Lavrent'ev, M. M., Romanov, V. G., and Shishatskii, S. P. (1980). *Ill-Posed Problems of Mathematical Physics and Analysis.* Nauka, Moscow (in Russian).

Liboff, R. (1979). *Introduction to the Theory of Kinetic Equations.* Krieger, Huntington.

Lions, J.-L. and Magenes, E. (1972). *Nonhomogeneous Boundary Value Problems and Applications.* Springer Verlag, Berlin - Heidelberg - London.

Ludwig, D. (1966). The Radon transform on Euclidean space. *Comm. Pure and Appl. Math.*, **19**.

Mikhailov, V. P. (1976). *Partial Differential Equations.* Nauka, Moscow (in Russian).

Morozov, V. A. (1974). *Regular Methods of Solving Ill-Posed Problems.* Moscow State University, Moscow (in Russian).

Mukhometov, R. G. (1975a). A certain problem of integral geometry. *Matematicheskie Problemy Geofiziki*, , No. 6, P. II, 212–242, Novosibirsk (in Russian).

Mukhometov, R. G. (1975b). Planar inverse kinematic problem of seismology. *Matematicheskie Problemy Geofiziki*, , No. 6, II, 243–252, Novosibirsk (in Russian).

Mukhometov, R. G. (1977). A problem of reconstruction of two-dimensional Riemannian metric, and integral geometry. *Dokl. Akad. Nauk SSSR*, **232**, No. 1, 32–35 (in Russian).

Mukhometov, R. G. (1981). On a problem of reconstructing Riemannian metrics. *Sibirsk. Mat. Zh.*, **22**, No. 3, 119–135 (in Russian).

Mukhometov, R. G. and Romanov, V. G. (1978). On the problem of finding an isotropic Riemannian metric in an n-dimensional space. *Dokl. Akad. Nauk SSSR*, **243**, No. 1, 41–44 (in Russian).

Naimark, M. A. (1967). *Linear Differential Operators.* F. Ungar, New York.

Natterer, F. (1986). *The Mathematics of Computerized Tomography.* John Wiley & Sons, New York.

Nirenberg, L. (1974). *Topics in Nonlinear Functional Analysis.* Lecture Notes, Courant Inst., New York Univ., New York.

Ovsyannikov, L. V. (1971). The nonlinear Cauchy problem in a scale of Banach spaces. *Dokl. Akad. Nauk SSSR*, **200**, No. 4, 789–792 (in Russian).

Pestov, L. N. and Sharafutdinov, V. A. (1988). Integral geometry of tensor fields on a manifold of negative curvature. *Soviet Math. Dokl.*, **36**, No. 1, 203–204.

Petrovskii, I. G. (1964). *Lectures on the Theory of Ordinary Differential Equations.* Nauka, Moscow (in Russian).

Plaksin, G. I. (1966a). Expressing a function through its integrals over ellipsoids. *Dokl. Akad. Nauk SSSR*, **166**, No. 3, 548–550 (in Russian).

Plaksin, G. I. (1966b). On a problem of I. M. Gel'fand. *Dokl. Akad. Nauk SSSR*, **170**, No. 4, 783–785 (in Russian).

Polyachenko, V. L. and Fridman, A. M. (1976). *Equilibrium and Stability in Gravitational Systems.* Nauka, Moscow (in Russian).

Prilepko, A. I. and Ivankov, A. L. (1985). Inverse problems of the determination of the coefficient and right-hand side of a nonstationary multivelocity transport equation from overdeterminedness at a point. *Differents. Uravn.* **21**, No. 1, 109–119 (in Russian).

Prilepko, A. I. and Volkov, N. P. (1987). Inverse problems for the determination of parameters of a nonstationary transport equation by overdeterminations of integral type. *Differents. Uravn.* **23**, No. 1, 124–136 (in Russian).

Radon, J. (1917). Über die Bestimmung vor Funktionen durch ihre Integralwarte längs gewisser Männigfritigkeiten. *Ber. Verh. Sächs Akad.*, , No. 69, 262–277.

Reznitskaya, K. G. (1975a). Connection between the solutions of the Cauchy problem for equations of different kinds and inverse problems. *Matematicheskie Problemy Geofiziki*, , No. 5, I, 55–62, Novosibirsk (in Russian).

Romanov, V. G. (1972). *Certain Inverse Problems for Equations of Hyperbolic Type.* Nauka, Novosibirsk (in Russian).

Romanov, V. G. (1978). Integral geometry on the geodesics of an isotropic Riemannian metric. *Dokl. Akad. Nauk SSSR*, **241**, No. 2, 290–293 (in Russian).

Romanov, V. G. (1984). *Inverse Problems of Mathematical Physics.* Nauka, Moscow (in Russian).

Romanov, V. G., Kabanikhin, S. I., and Boboev, K. (1983). An inverse problem for P_n-approximation of the kinetic transport equation. *Dokl. Akad. Nauk SSSR*, **276**, No. 2, 296–299 (in Russian).

Samarskii, A. A. (1980). Some problems of the theory of differential equations. *Differents. Uravn.* **16**, No. 11, 1925–1935 (in Russian).

Santalo, L. (1976). *Integral Geometry and Geometric Probability.* Addison-Wesley Publishing Co., Reading.

Sharafutdinov, V. A. (1981). Determining the characteristics of an optical body in a homogeneous medium from its images. In: *Mathematical Methods of Solving Direct and Inverse Problems of Geophysics*, 123–148, Novosibirsk (in Russian).

Sharafutdinov, V. A. (1993). A modified horizontal derivative and some of its applications. *Siberian Math. J.*, **36**, No. 3, 574–607.

Sharafutdinov, V. A. (1994a). *Integral Geometry of Tensor Fields.* VSP, Utrecht.

Sharafutdinov, V. A. (1994b). Mathematical aspects of polarization tomography. In: *Applications of the Radon Transform*, 120–140, Amer. Math. Soc., Princeton.

Silin, V. P. (1971). *Introduction to the Kinetic Theory of Gases.* Nauka, Moscow (in Russian).

Smirnov, V. I. (1969). *A Course in Higher Mathematics.* Nauka, Moscow (in Russian).

Tanana, V. P. (1997). *Methods for Solving Operator Equations.* VSP, Utrecht

Taylor, M. (1981). *Pseudodifferential Operators.* Princeton University Press, Princeton.

Tikhonov, A. N. and Arsenin, V. Ya. (1979). *Methods of Solution of Ill-Posed Problems.* Nauka, Moscow (in Russian).

Tikhonov, A. N., Arsenin, V. Ya., and Timonov A. A. (1987). *Mathematical Problems of Computerized Tomography.* Nauka, Moscow (in Russian).

Uspenskii, S. V. (1972). Reconstruction of a function specified by integrals over a family of ellipsoids. *Sibirsk. Mat. Zh.*, **13**, No. 6, 1374–1382.

Uspenskii, S. V. (1977). The reconstruction of a function given by integrals over a family of conical surfaces. *Sibirsk. Mat. Zh.*, **18**, No. 3, 675–684 (in Russian).

Vainshtein, B. K. (1973). Three-dimensional electronic microscopy of biological macromolecules. *Uspekhi Fiz. Nauk*, **109**, No. 3 (in Russian).

Vekua, I. N. (1959). *Generalized Analytic Functions.* Nauka, Moscow (in Russian) (English transl.: *Generalized Analytic Functions*, 1962, Addison-Wesley Publishing Co., Reading.)

Vladimirov, V. S. (1961). Mathematical problems of the uniform-speed theory of transport. In: *Trudy Mat. Inst. Steklov*, **61** (in Russian).

Vladimirov, V. S. and Volovich, I. V. (1985). Local and nonlocal currents for nonlinear equations. *Teoret. Mat. Fiz.*, **62**, No. 1, 3–29 (in Russian).

Yakhno, V. G. (1982). Theorem of existence in the small for a certain multidimensional problem. In: *Problems of Well-Posedness of Inverse Problems of Mathematical Physics.* 143–155, Novosibirsk (in Russian).

Bei Fragen zur Produktsicherheit wenden Sie sich bitte an:
If you have any questions regarding product safety,
please contact:

Walter de Gruyter GmbH
Genthiner Straße 13
10785 Berlin
productsafety@degruyterbrill.com